Praise for *The Accidental Seed Heroes*

'There can be few tasks more important than offering us a glimpse at how our broken food system might be reimagined, from the ground up; as ever, Adam does this brilliantly. A special, important book of hope, action and integrity.'

MARK DIACONO, food and garden writer

'This is a fascinating insight into the world of breeding, growing and the preservation of endangered fruit, vegetables and cereals around the world. Adam takes us to far-flung places, and we are introduced to some remarkable people whose passion for their own local food culture champions traditional varieties while also breeding new ones. In his search for 'deliciousness' in the gardens and farms of tiny villages, it is possible to see how such underutilised indigenous produce can contribute to a more biologically diverse future in feeding us all. I loved this book and enjoyed the nice, short, easily digestible chapters, perfect for dipping into at your leisure – although I felt compelled to keep reading!'

ADVOLLY RICHMOND, plant and garden historian;
author of *A Short History of Flowers*

'If you're a fan of growing your own food or just curious about where your veggies come from, *The Accidental Seed Heroes* is a cracking read. Adam weaves together engaging stories about traditional farmers (the real unsung heroes) who've helped preserve the diversity of crops we often take for granted. It's a reminder of how vital this diversity is, especially as we face the challenges of a changing climate.

'What I love most is how it bridges the past and present. Alongside the tales of heritage farming, Adam highlights modern plant-breeding innovations that focus on sustainability and flavour, without relying on a cocktail of chemicals. It feels incredibly relevant for gardeners today, whether you're pottering about in a small urban patch or tending an allotment like mine.

'This isn't just a book about seeds or food systems; it's an inspiring nudge to get stuck into the joys of gardening while understanding the bigger picture. A proper gem for anyone keen to connect their love of growing with the wider world of growing.'

ROB SMITH, author of *Grow to Eat*

'Adam explores plant diversity in food production, past and present, and scrutinises future implications for food security. He skilfully distils a complex subject into clear, accessible concepts, travelling through countries and meeting scientists to untangle the political and commercial influences at play. Educational, inspiring and motivational, *The Accidental Seed Heroes* illustrates how, as individual gardeners, we have a role to play in fostering both local and global resilience in the food supply system.'

SUE KENT, TV presenter; author of *Garden Notes*

'After reading Adam's book, I won't look at a handful of seed the same way again! A fascinating and enlightening read, beautifully told. *The Accidental Seed Heroes* delves into the history of breeding and how we ended up with seed superpowers dominating what we grow and eat. It then takes us on a journey of sustainability, resourcefulness, resilience and the expertise of the growers at the cutting edge. With everything that's currently going on in the world politically and environmentally, this book feels particularly relevant.'

JOE SWIFT, garden designer;
writer; presenter, BBC's *Gardeners' World*

'From the moment you turn the first page, Adam invites you on a spellbinding journey around the globe, offering a unique perspective on food security and the critical role of seed saving. The book takes readers from the arid plains of Rajasthan to the lush landscapes of Ethiopia, introducing the passionate and inspiring individuals working tirelessly to safeguard the future of our food supply through the preservation and cultivation of diverse plant varieties.

'The narrative is immersive, and Adam's writing style imbues the scientific topic with a gentle, almost romantic air. If you're someone who has never considered saving seeds before, Adam will undoubtedly change your perspective.

'Overall, this book is a thought-provoking exploration that combines science, history and travel, leaving the reader with a renewed appreciation for the small yet powerful act of saving seeds. Whether you're an avid gardener or simply someone concerned with the future of food, *The Accidental Seed Heroes* offers a captivating and enlightening journey into the heart of agricultural preservation.'

AUGUST BERNSTEIN,
head tutor, Raymond Blanc Gardening School

'What a timely and fascinating book this is. Adam Alexander has created an extremely important work that adds to the library of not only plant breeders but all gardeners and farmers. In truth, everyone should seek a deeper understanding of the origins of the foods we eat and the challenges to be faced in the future, and this book perfectly serves that purpose.'

CRAIG LEHOULLIER,
author of *Epic Tomatoes*; co-lead, Dwarf Tomato Project

'In this delicious travelogue, you will encounter memorable varieties with engaging names like Big Gogozahare sweet pepper, Honest Ærling's Genuine Eating Pea and Awesome Emma tomato, and hear gripping stories about how they originate. You'll meet the dedicated, passionate, unforgettable people who create and tend them, discover unfamiliar crops, such as enset, khat and maslins, and dishes from them like *boula* porridge, and learn techniques of crop diversification handed down from indigenous cultures all over the world, over countless generations. Those interested in enriching their cuisine will draw inspiration from Adam's journeys and discoveries. Those drawn to the seed arts (growing, saving, selecting, multiplying and breeding) will be fascinated by the variety of available techniques, from the ancient wisdom of indigenous cultures to the most modern genomics and how the complex interactions between these varied models are impacting our food supply.'

C.R. LAWN, founder, Fedco Seeds

'This is a really enjoyable book that should appeal to anyone who eats. Adam Alexander does a great job of explaining the problems with our current food system and how it's providing us with bland, identikit vegetables and cereals. But, more importantly, he introduces and celebrates people across the world who are stewarding and developing the seeds that could form the basis of a better, fairer and tastier way to eat for all of us.'

KATE MCEVOY, co-founder, Real Seeds

'In this riveting book, Adam Alexander points to the glories, creativity, fun and failures of the world's plant breeders. We meet thousands of seed Davids sidestepping the seed Goliaths. We salute scientists who believe seeds are public, not private, property. We go to highlands and lowlands in countless countries meeting the frontline seed savers, unsung heroes of diversity, on whom our future food depends. His erudition and excitement are palpable as he shares what they, we and he can grow and develop. This is a hymn to the joys of planting and developing seeds for the common good. You'll look at seed packets differently after reading this book.'

TIM LANG, professor emeritus of food policy,
City St George's, University of London

'There is something about popular science books that can enchant me, and this one certainly does. The mixture of quite complicated concepts and real-life experiences is captivating. It is wonderful to be shown by Adam the importance of seed, local varieties and some of my favourite vegetables (and grains), and to be taken on stories from across the world and his own back garden. A beautiful book telling an important story.'

BRUCE PEARCE, director of science, Garden Organic

THE
ACCIDENTAL
SEED
HEROES

THE ACCIDENTAL SEED HEROES

Growing a delicious
food future for all of us

Adam Alexander

Foreword by Rekha Mistry

Chelsea Green Publishing
London, UK
White River Junction, Vermont, USA

First published in 2025 by Chelsea Green Publishing | PO Box 4529 | White River Junction, VT 05001 | West Wing, Somerset House, Strand | London, WC2R 1LA, UK | www.chelseagreen.com
A Division of Rizzoli International Publications, Inc. | 49 West 27th Street | New York, NY 10001 | www.rizzoliusa.com

Copyright © 2025 by Adam Alexander.
All rights reserved.

No part of this book may be transmitted or reproduced in any form by any means without permission in writing from the publisher.

Publisher: Charles Miers
Deputy Publisher: Matthew Derr
Commissioning Editor: Muna Reyal
Project Manager: Susan Pegg
Copy Editor: Susan Pegg
Proofreader: Jacqui Lewis
Indexer: Charmian Parkin
Designer: Melissa Jacobson
Page Layout: Jenna Richardson

ISBN 978-1-915294-43-2 (hardback) | ISBN 978-1-64502-364-7 (ebook) | ISBN 978-1-915294-44-9 (UK ebook) | ISBN 978-1-915294-45-6 (audiobook)
Library of Congress Control Number: 2025001018 (print)
A CIP catalogue record for this book is available from the British Library.

Our Commitment to Green Publishing
Chelsea Green sees publishing as a tool for cultural change and ecological stewardship. We strive to align our book manufacturing practices with our editorial mission and to reduce the impact of our business enterprise in the environment. We print our books on chlorine-free recycled paper, using vegetable-based inks whenever possible. This book may cost slightly more because it was printed on paper that contains recycled fiber, and we hope you'll agree that it's worth it. *The Accidental Seed Heroes* was printed on paper supplied by Sheridan that is made of recycled materials and other controlled sources.

Authorized EU representative for product safety and compliance
Mondadori Libri S.p.A. | www.mondadori.it
via Gian Battista Vico 42 | Milan, Italy 20123

Printed in the United States of America.
10 9 8 7 6 5 4 3 2 1   25 26 27 28 29

For Julia

Contents

FOREWORD BY REKHA MISTRY — xiii

Introduction — 1

1: Breaking the Mould
A Brief History of Plant Breeding — 11

2: Where Farmers' Varieties Reign Supreme
On the Trail of Deliciousness in Albania and Ethiopia — 31

3: Using One's Loaf
The Bigger the Population, the Better — 51

4: Setting Seeds Free
Plant Breeding for an Equitable Planet — 75

5: A Future Full of Beans
A Solution to Save the World? — 93

6: Cultivating Capsicums
A Tasty Future — 114

7: Red Is Not the Only Colour
The Quest for Deliciousness in Every Bite — 131

8: Perfecting the Perfect Pea
Sometimes Uniformity Can Be a Good Thing — 151

9: Let Us Eat Leaves
And Other Bitter Beauties — 173

10: Beautiful Brinjal
The Making of an Asian Love Affair — 194

11: It's All in the Pip
Fruitful Labour for Apple Breeders — 211

Conclusion: Holding Truth to Power — 233

ACKNOWLEDGEMENTS — 245
GLOSSARY — 248
NOTES — 255
INDEX — 271

Foreword

Seed saving is so precious to me. I've been championing saving seeds since I took on an allotment in 2011. At the time, seed saving was frowned upon. I was taken aback. To these veteran plot holders, it looked like I put a lot of effort into saving a handful of seeds. How to change generational mindset? Is it really that much effort?

When I first saved flower and vegetable seeds from our garden over thirty years ago, it was done without thought; you could say it was, quite simply, second nature. I was born and brought up in Lusaka, Zambia. My mother used to tend her vegetable garden daily. At the time, I missed something. As a young child and through my teens, I'd help Mum in the garden, but I would watch her and not see the absolute joy in her face as she wistfully walked along an avenue of 2-metre-tall pigeon pea plants that she had started from saved seed from our ancestral family garden in India. Her beady eyes always kept an eye over the aubergine and tomato fruits hidden among – or rather *protected by* – the abundant rows of lustrous marigold flower shades. One memory I fondly recall is her success after she experimented growing ginger and turmeric for a few years in Lusaka's savanna climate. Witnessing an impressive harvest of ginger rhizomes from no more than a metre square in that small courtyard space is a moment that triggered my own interest in gardening and one that will always remain a precious memory.

The Accidental Seed Heroes

Only when I started my own allotment garden did I understand my mother's sense of pride. I am so proud of what she instilled in me. She saw an unkept patch of land outside her own small flower garden and turned it into a vegetable heaven. Mum always saved the seeds of what she grew. Even when the crop was unsuccessful, she saved a few seeds from that harvest. That was essential work: saving folk variety (farmers' variety) seeds. She would always say it was her link to her heritage.

After living in Lusaka for over forty years, my parents retired back to Gujarat, India, and moved into an apartment. But that didn't stop my mother – she took her collection of saved seed with her. Walk onto their balcony and you'll not only find her beloved holy basil plants but also her experimental vegetables. Whenever I visit my parents, her garden always comes up and her eyes light up. She reminds me of why she would ask us to help her in the garden. How she used to love experimenting, trying to grow the vegetables she grew up with in India or had simply found in the local markets. She saved, swapped, shared and collected seeds! Gardening was – is – her passion, and I am now living it for her. I loved our garden in Lusaka. It was my thinking space. It became my escape. Little did I know her gardening ethos would rub off on me in the way that it has. I wouldn't be where I stand today! Enough said. I am her daughter!

In Adam's first book, *The Seed Detective*, he brought our attention to heritage seed varieties and their value. In this book, prepare yourselves to understand that it's not just the forgotten art of seed saving but the loss, or the near loss, of many heritage seeds that we need to be concerned about. But are they lost? Throughout this book, Adam's thorough

xiv

Foreword

research and findings has shone a light on some incredible individuals who have been saving seed and, quite frankly, I believe it will help us understand our heritage seed. We are a great nation that has jumped onto the 'save the pollinators' wagon. The message is wide and, more importantly, the message is clear. We have changed our practices and banned chemicals to help them thrive. With the same gusto, how about we all climb onto the 'save the seeds' wagon? If we all take some small seed-saving steps, we will bring some giant changes for food crops.

Using One's Loaf, Adam's chapter on bread and its many guises, struck a huge chord with me. A simple grass or grain that started life in the Fertile Crescent (see Wheat and Us on page 54). Throughout this chapter, I kept thinking of where the next batch of flour for my chapati or bread rolls will come from. Modern cultivars of wheat are seen by commercial players, governments and big business as the answer to the world's hungry population – the twin goals of intensive farming and maximising profits mean that farmers are pressured into choosing these specifically bred seeds. Thankfully, as Adam discovers, there are *heroes* out there who are saving the last remaining open-pollinated and tasty folk varieties. We learn, thanks to Adam for sharing this, their motivation: it's not about money or mass production but quality and heritage. This couldn't be more true. We are what we eat! Words are not enough; this chapter has genuinely moved me and how I see all grains. Thank you, Seed Heroes.

I've not told Adam this, but he is my accidental seed hero! I first came across his work when I joined the Heritage Seed Library over a decade ago. Since then, he continues to inspire me. Not only to save seeds but to enjoy the taste of these

The Accidental Seed Heroes

so-called forgotten crops. To champion these heritage varieties and folk varieties, and to make them even more relevant in today's fast-grown food fields. The way Adam champions seed-saving work gave *me* the strength and bravery I needed not to feel afraid to champion seed saving, whatever the genus or variety.

Thank you, too, to everyone mentioned within the book. Every chapter strikes a cultural, ethical and sustainable message that we must all take note of. Thank you, the Accidental Seed Heroes, I applaud you all.

REKHA MISTRY
Head Gardener at Haddon Hall,
TV presenter and author of *Rekha's Kitchen Garden*

Introduction

Nature, ere she gives up, makes a violent effort to reproduce.

Isaac Anderson-Henry, *Transactions of the Botanical Society* (1867)

The landscape was green, the terraces sculpturing the rolling hills and mountains with their human imprint. At 2,000 metres elevation and 5 degrees north of the equator, the air was warm and fresh from recent rains – the grey clouds, like brooding monsters, scuttling across the sky. In that moment, a more perfect climate for growing crops seemed hard to imagine. On this convivial Sunday morning my lovely driver, Ermias, had dropped me and local guide Genale Geyato at his village, Mecheke. It nestles in the centre of a UNESCO World Heritage Site, the Konso Cultural Landscape in southern Ethiopia, and what this most beautiful of regions has to teach us all about sustainable agriculture and plant breeding cannot be overestimated. I had come to this remote corner of the country to see how traditional agroforestry is practised and how the genius and hard work of the farmers have ensured a resilient and stable food supply for the last eight hundred years.

The indigenous people of this region maintain and cultivate their crops in a land of terraces. In all of those eight centuries, despite the fact that the soil is thin and requires constant husbanding, Genale told me they have never once suffered from famine, nor have they had their land degraded by erosion due

The Accidental Seed Heroes

to drought or flood. As we strolled along the narrow paths and among the round mud and thatch houses of Genale's tribe, an old lady carrying a bouquet of deep purple sorghum seed heads slung over her shoulder – destined to be turned into a refreshing and tasty alcoholic beer I was soon to savour – walked past us as a gaggle of curious kids gathered around. What the farmers tending to their terraces told me about maintaining the genetic diversity of their crops, about living in a world of food insecurity where the smallest deviation in the weather – rain coming too early or too late, or there being too much or too little of it – could prove fatal, left me humbled and in awe. As with other indigenous farmers around the world, I sit at their feet. That admiration, coupled with a curiosity to understand more about the choices we have in how we develop and maintain the crops that will nourish all of us as our climate changes, and a wish to acknowledge their wisdom and skill, has led me to write this book. The farmers in this region have been selecting, improving, maintaining and celebrating the seeds of their harvests for centuries; theirs is just one example of how different approaches to plant breeding offer more hopeful routes to a future where we can feed ourselves, and not at the planet's expense.

In my previous book, *The Seed Detective*, I told the stories of many vegetables' journeys from wild parent to cultivated offspring and their place as history on a plate. In this book, I don my seed detective homburg once more to uncover the remarkable stories and check out the flavours of a new generation of crops that exist because of a passionate and committed cohort of breeders and growers, both traditional and modern, around the world. They are showing the way forward, not only championing traditional varieties, but breeding delicious new ones that are fundamental to a sustainable future for our planet. Meeting these people and savouring the delights of

Introduction

their creations, I am filled with a sense of optimism. We can and do breed crops that will feed us as our climate becomes ever more extreme and unpredictable – and that are not dependent on chemical inputs, monoculture and uniformity. Maintaining and improving traditional crops and breeding new cultivars* that do best in low-input growing systems – ones that require little or no use of chemical fertilisers in order to flourish – and that are seen as a public good should be our mantra. They should stay firmly in the hands of indigenous farmers, independent local breeders, and eccentric, obsessive and passionately committed amateurs and professionals who believe in freely sharing their work.[†]

The world needs this essential counter to the hegemony and hubris of a globalised and commodified system because, since the middle of the twentieth century, plant breeding has been focused on creating cultivars that are designed to deliver greater yields within a system of monoculture. This has created an existential threat for us all because this system has bred out diversity and resilience within our most important crops, with

---

* The term cultivar – 'cultivated variety' – describes a plant that is the result of selective breeding between two known parents; most commercial varieties bred in the last one hundred years are cultivars. This is different from a variety, which arises as the result of a non-human induced crossing, accidental pollination or mutation. Despite this distinction, the two terms are often used interchangeably.

† I use various terms to describe types of amateur and professional breeders throughout the book. Open-source breeding is an approach that allows and encourages other breeders to use new cultivars in their own breeding programmes without recourse to patent protection or licensing. Freelance breeder describes anyone working independently outside commercial and mainstream organisations.

The Accidental Seed Heroes

the result that they become ever more susceptible to catastrophic failures due to climate change, pests and diseases. As we shall see, embracing a diverse and holistic approach to breeding is fundamental to ensuring nutritious and plentiful harvests in a changing climate, drawing on the best of science, invention, curiosity and our deep, past knowledge. Not to mention the pursuit of deliciousness. The characters that feature throughout this book – the crops and the people who nurture them – inspire solutions to breaking the current model, in which seeds are considered to be intellectual property, controlled by patents and owned by a handful of giant agribusiness monopolies. I cherish seed as a common resource that all the world should be able to access freely. Seeds reinforce our diverse cultural identities, and a celebration of the place of plant breeding, in all its forms, lies within the stories in the following pages.

In talking about different cultures and parts of the world, one of the challenges has been to avoid lumping entire regions into binary categories in order to describe their approaches and philosophies towards breeding and maintaining the crops we rely on to survive. Terms like the West and the East, the developed or developing world and others, which some might consider pejorative, are all describing economic activity and status. Over the years, they have created much noise, misunderstanding and debate. Although they are far from being perfect because of their use within an economic context, I have chosen to use the terms Global South and Global North to describe a particular reality: where homogenous monoculture farming is practised extensively – but not exclusively – in the wealthiest countries of the northern hemisphere and the antithesis of this approach is seen to be employed more extensively – but also not exclusively – in the southern hemisphere. I do not use these terms as economic indicators.

Introduction

A Changing Climate in Plant Breeding

'You're an accidental plant breeder whether you think so or not!' said Carol Deppe, the godmother of American plant breeders and author of the plant breeder's bible, *Breed Your Own Vegetable Varieties*.[1] I had been in touch with her as I embarked on writing this book because plant breeding was a subject I knew virtually nothing about. I was keen to understand something about the work of the Open Source Seed Initiative (OSSI), which was founded in the US in 2012. At the time of writing, Carol is its chair, and their mission is 'maintaining fair and open access to plant genetic resources worldwide in order to ensure the availability of germplasm [seeds] to farmers, gardeners, breeders, and communities of this and future generations'.[2]

Up till this point, in the thirty-five years I had spent collecting, saving and sharing seeds, I had presumed all I was doing was maintaining varieties, ensuring that the seed I saved was the same as the seed I had sown. I hadn't thought that continually selecting seeds from the first fruits and pods to ripen was a form of plant breeding; that varieties I had been saving over many generations and that were very happy growing in my corner of South Wales had become locally adapted – so-called landraces, or farmers' or folk varieties (FVs).* According to Carol, I now qualify as a backyard breeder because the decisions I make about the seeds I save mean my crops change too.

---

* A landrace, or farmers' or folk variety (FV), is defined as a plant population with a limited range of genetic variations, which is adapted to local agroclimatic conditions and which has been generated, selected, named and maintained by traditional and indigenous farmers. These terms are more or less synonymous; for simplicity's sake I have used FV throughout this book.

The Accidental Seed Heroes

Since the middle of the twentieth century, a combination of highly mechanised farming, continuous cropping with monocultures and increased dependency on high-yielding homogeneous cultivars of the three most important crops in the world – wheat, rice and maize – has resulted in a 90 per cent reduction in the genetic diversity of our food. Just five giant agribusinesses breed and sell 40 per cent of all the seed in the world. Their business models perpetuate the use of monoculture as *the* solution to producing more food at the expense both of greater diversity and innovation and of small, local and highly adaptive breeding. Dependency on homogeneous, genetically narrow cultivars that are grown as monocrops is a high-risk strategy. A single pathogen can wipe out swathes of crops because all the plants are identical and any chink in their genetic armour is easily exploited. A warming world is creating an ever more benign environment for new and deadly pathogens and pests to evolve. As I hope to show, although there is a place for this form of modern food production in certain places, as a model it needs to be replaced with a suite of solutions that ensure innovation puts people and planet first.

So, as we enter the second quarter of the twenty-first century, can we realistically look forward to a point in the next twenty-five years where plant breeding is no longer a monopoly? Is the tide turning? I believe there are compelling reasons to be hopeful. Indigenous and traditional farmers from the Global South, hobby farmers, academic institutions, amateur and professional breeders with an emphasis on seeds best suited to organic growing, frequently working together, are maintaining, developing and breeding new cultivars that are fit for purpose: crops that are nutritionally dense, delicious, need less water and can cope in more extreme

Introduction

climatic conditions. Driven by altruism and collaboration, these people are working with wild relatives, local varieties, FVs and older commercial varieties. For the most part, their new varieties are open-pollinated – pollinated by wind or animals – and 'breed true', meaning the progeny are identical to the parent and the grower knows the seeds in the packet are what the packet says they are! Using a combination of traditional methods and technologies, alongside a diverse number of approaches and local solutions to restore food security – the ability to feed ourselves in the face of regional or global conflict, weather disasters and with less dependence on a globalised food system – means we really can improve not only our health but the planet's too.

In the following pages, I look at the diverse approaches to plant breeding that are employed across many different species and types of crop. They fall into two distinct types. The first is phenotype breeding, which is based on considering a plant's morphology – its observable characteristics – transferring pollen from a male flower to a female one (usually with a small brush), a technique first used by botanists and scientists three hundred years ago.* Phenotype breeding also includes mutagenesis, the forced mutation of a plant using either chemicals or radiation, a popular and successful breeding method that has been in place for at least one hundred years.

The second type is genotype or molecular breeding. This includes transgenics: the transfer of genes from one species into another, which is the basis of all forms of genetic modification. This is different from genetic engineering or

---

* This type of breeding can be variously described as conventional, classical or traditional plant breeding; I use all three terms throughout the book and consider them fully interchangeable.

The Accidental Seed Heroes

genome editing, which is the transfer or removal of genes between individual plants of the same species. In the same way that when I am fertilising a courgette flower I use a paintbrush to transfer pollen from a male to a female plant, geneticists (modern plant scientists) use the equivalent of a pair of scissors or a scalpel to edit genes within an organism's DNA. Another tool is marker-assisted selection, also known as marker-aided selection or MAS. This is a technology that looks at DNA-based genetic markers within a plant's genome – its genetic barcode – and identifies those that are associated with specific traits the breeder wants to include in their new cultivar. MAS has, until now, been used almost exclusively by geneticists. However, with genetic sequencing facilities at universities being made more widely available, conventional breeders are starting to use this amazing tool to speed up selections of varieties employing classical phenotype breeding. It is democratising plant breeding in the Global South, especially with so-called 'orphan crops', which feature in many of the following chapters. *

We don't all need to become backyard breeders or even, like me, accidental ones; as growers, we don't even need to eschew many of the modern hybrid cultivars our seed catalogues are stuffed with. As citizens I am not suggesting we boycott those same uninspiring specimens that populate our supermarket shelves – though you won't catch me buying them. After all, it should be first and foremost a matter of personal choice what we decide to grow and eat. I just want that choice to be better informed and infinitely more diverse

---

* Neglected and underutilised indigenous crops, usually grown by subsistence farmers, are described as orphan crops by plant scientists and ethnobotanists.

Introduction

and enjoyable. But will new, independently developed strains of fruits, vegetables and grains sustain us all, not just with nutritious and delicious food, but as part of the solution to combating climate change and returning fertility to our soils and biodiversity to our land? I passionately believe they will.

My quest to find and savour the crops that can and must be part of building diverse, resilient and nature-friendly solutions to feed the world has taken me to Rajasthan, the US, my own backyard in Wales, across Europe to the remote regions of southern Albania, and into the southern part of the Great Rift Valley of Ethiopia. Throughout my travels, I have come face to face with crops as exotic as they can be prosaic: lettuce, peppers and chillies, aubergines, wheat, onions, beans, peas and tomatoes. Also, indigenous crops, such as enset, tef and sorghum – staples for millions in the Global South – and traditional cereal mixes known as maslins, which provide resilience and a harvest in the face of extreme climate events. Finally, I include one of the most important and delicious of fruits that exemplifies the challenges and complexities in different approaches to breeding: the apple.

A Taste for an Identity

'What do you recommend I grow?' It's the one question above all others that people ask me. Impossible to answer without knowing something about the poser of the question. My first response is always, 'What do you like to eat?' This is as much to tease out what matters to them on both a culinary and a cultural level as to understand the circumstances under which they are planning to breed, grow and consume their beloved crops. This matters because feeling a connection to our food – its provenance, its place in our own stories and identity, its

The Accidental Seed Heroes

flavours and uses – becomes the starting point on a journey where we care. I care very much because I feel connected to what I grow and eat. I believe that the route to a healthier world is to celebrate foods that are local to us, that enhance biodiversity and have natural resilience to pests and diseases: crops that can evolve and cope with the inevitable extremes of a changing climate. All of us who grow vegetables and save seeds are part of the solution, whether we like it or not!

I feel empowered by my association and connection with varieties that matter to me; not only because they benefit me nutritionally and taste great, but because they connect me to a positive journey towards a more sustainable and healthy world. It goes without saying that it all starts with seeds – those that evolve and adapt as we and our world do. Seeds strengthen our connections to what we grow and eat; they are intrinsic to our identity and our future. So, the first questions I ask those you will meet in the following pages are, 'Is it delicious?' and 'Why is it important to you?' Two things that we might also ask ourselves and that have led me on a journey to discover how many exciting, inspiring and empowering people there are out there; people who offer hope and insights into a future for our world that is rich with flavour and amazing foods: a solution to a carbon-neutral planet by mid-century. I hope that sharing this journey with me will instil in you the same feelings of hope I have that great things in the field of food production are already making the world a better place for all living things.

CHAPTER ONE

Breaking the Mould
A Brief History of Plant Breeding

A process for breeding of plants or animals is essentially biological if it is exclusively based on natural phenomena such as crossing, selection, non-targeted mutagenesis or random genetic variations that occur in nature.

Definition of classical breeding
under Austrian law

Until the middle of the nineteenth century, farmers around the world were doing two jobs: growing crops and working as plant breeders. Not in the way we currently might think of breeding, by deliberately crossing different varieties of the same species, but through a process of selection through observation. Today, this tradition continues in many parts of the world, especially the Global South. But for the Global North this dual role has been long forgotten, thanks to a revolution in plant breeding that began about 150 years ago.

From a Common Good to the Property of a Multinational

Seeds have been shared, traded and continuously selected for traits the farmers want since the dawn of settled agriculture about twelve thousand years ago. This ensured that local adaptation was an ongoing process, producing seed that could flourish in a fluid climatic environment. Even as the climate heats up and we experience ever greater and more frequent extremes of weather, having locally adapted varieties ensures not only greater food security but also an environment of cultivation that goes hand in hand with evolving valuable changes in our crops. As long as farmers were plant breeders as well as food producers, seeds were considered as 'commons' – public property, of benefit to all. Their place in the economy of agriculture was recognised as a public good. It's very different now. Not only are the major crops that feed us treated as commodities, assets of global trade, but seeds are too.

Our attitude to plant breeding started to change two hundred years ago. By the end of the first quarter of the nineteenth century, a new generation of 'natural scientists' had emerged. They were botanists and the first plant scientists to apply a systematic approach to breeding; one intimately bound up with improvements in farming technology and industrialisation.

A Missed Opportunity

Britain was the first country to start a fully scientific approach to plant breeding. It began in the early 1890s when Dr John Garton (1863–1922), who had worked in the family business as a seed merchant, set up R. & J. Garton with his brother, at Newton-le-Willows in Lancashire. He understood that some plants were fertilised through cross-pollinating and others by self-pollinating. With this knowledge, he began an extensive breeding programme, systematically crossing thousands of combinations of different varieties of cereals on the family farm. His first success was an oat he named

Breaking the Mould

Abundance, which went on sale to farmers in 1892. This new approach, which Garton called 'scientific farm plant breeding', caught the attention of academics and the Royal Agricultural Societies of England and Scotland. They understood that his methodology had the potential to create an endless number of new and distinct varieties of cereals.

From the start, John Garton believed that what he was doing was for the public good. He created a great diversity of new varieties, which were grown across the British Empire and that he originally hoped could be distributed for free. His work was of great interest to the British government, not only because security of cereal production was critical but also because his speciality, oats, were considered of international importance to the British Empire and also the US. The fact that his research was not dependent on public funds also helped perpetuate a narrative that governments needn't always fund development of these major crops. Nonetheless, the British government established the Board of Agriculture in 1889 and, in exchange for the new department supporting his research and breeding programme, Garton offered to give them his entire body of work, so long as they would make seeds available to farmers, preferably free of charge, otherwise at market prices.

Despite enthusiastic support and endorsements from fellow agronomists, botanists and research scientists, who hoped the government could see the benefits to the Empire, the Board turned down Garton's offer, saying there was no precedent for this model of public and private partnership. In his introduction to his company's 1899 spring catalogue, Garton wrote:

> Our efforts in this direction not having been successful, and as we were not in a position to

> undertake the work of distribution ourselves, we
> have placed [R.&J. Garton] in the hands of a Public
> Company, and we trust that the continued efforts
> made by us on behalf of the British farmer will
> be fully appreciated by him, through his support
> of the Company responsible for the distribution of
> the seed of our new breeds of agricultural plants.

In 1898, the family business went public with many investors being farmers, and Gartons Limited was born. It was listed on the London Stock Exchange in 1947, and folded in 1983.

The American Influence

Early in the nineteenth century, the constitution of the Columbian Institute for the Promotion of Arts and Sciences in Washington DC proposed the creation of a botanic garden to collect, grow and distribute plants of North America and countries around the world that might contribute to the welfare of the American people. In 1842, the United States Exploring Expedition to the South Seas (otherwise known as the Wilkes Expedition) brought back a collection of living plants. It formed the first permanent collection for the United States Botanic Garden in Washington and seeded the idea of an institution that was to be of fundamental importance to the development of plant breeding: the United States Department of Agriculture (USDA). It was charged with delivering a national programme of seed distribution to farmers for them to grow on and use as breeding stock.[1] In 1843, the secretary of war, James Porter, had pine boxes put on board all US ships to enable the collection of plants and seeds for the benefit of American farmers. For the first time, a government became involved in plant research and development. At the start of the American Civil War in 1861, there were greenhouses in Washington DC stuffed with East Asian plants – including fifty thousand tea plants (*Camellia sinensis*) and fruits like lychees, grapes, figs and pomegranates – all with a view to seeing how they could be added to the diversity of crops under cultivation in the US. Plant material and seeds from East Asia were much sought after because the area's climate was similar to that of the southern states of the US. Other crops of particular interest were soya beans and cereals such as sorghum and even opium poppies, which it was believed could become important cash crops for American farmers.

In order to kick-start a revolution in food production and the development of improved varieties, the US government provided seeds free of charge to a variety of growers including a new generation of wealthy horticulturalists – hobbyists and enthusiastic amateur plant breeders – with a particular interest in vegetables they could grow in the eastern states. Recently arrived immigrants from eastern Europe, who were transforming agriculture in the Midwest, were also important recipients of these seeds and were particularly keen to grow wheat varieties that they were familiar with from home. Many farmers and growers reported the results of growing these new imported varieties to the government, whether successful or otherwise, and there was always a demand for more seed. All the improvements and developments of new wheat varieties were thanks to farmers.

The advent of the Civil War seriously disrupted the provision of free seed. Many Americans objected to the policy, some seeing it as elitist and partisan – a reflection of the political and social divisions of the time. But probably the biggest change was in the mindset of the beneficiaries of US government policy: the new entrepreneurs among nurserymen and women and plant breeders who were trying to improve seed quality and diversity, which they could make a commercial success of. They accused the government of competing unfairly with them by funding plant innovation from the public purse. They believed the federal government had no business being involved in research and developments in plant breeding and seed production. Now seeds were seen as a product of commerce, a source of profit, rather than a shared resource for the common good – a ploy that was to be used again with further innovations in plant breeding.

Despite this prevailing view among plant breeders, the actions of the federal government led to an explosion in new seed businesses. Their creations were open-pollinated and there was no copyright on the products of plant breeding. Anyone could grow a crop, save the seeds and sell them or give them away if they so chose. This all changed early in the twentieth century when the greatest revolution in our understanding of plant breeding took place: the discovery of genetics.

Britain Takes the Lead

For the last thirty-five years of the nineteenth century, the work of the biologist and Augustinian friar Gregor Mendel (1822–1884), who described the rules of inheritance through his work with peas in Germany, had been ignored. His paper, 'Versuche über Pflanzen-Hybriden' ('Experiments on Plant Hybridisation'), published in 1865, remained largely unread until three independent researchers – the botanists Erich Tschermak from Austria, Carl Correns from Germany

Breaking the Mould

and Hugo de Vries from the Netherlands – published their responses in 1900. In the same year, the British biologist William Bateson (1861–1926) presented Mendel's paper to the Royal Horticultural Society (RHS) and organised an English translation, which was published the following year. Bateson admired Mendel's work, which was considered controversial at the time, writing a defence in 1902. In 1905, in response to a request to Cambridge University for the foundation of a professorship into the study of heredity, Bateson suggested a single word that would describe the work: genetics. The rules of inheritance that Mendel discovered would now be applied in a systematic way to develop entirely new crop varieties. Consequently, a new player entered the scene: the scientist. The farmer no longer held sway; instead, someone in a white coat working in a laboratory became the new plant breeder, moving ever further away from the farmer's field and onto the trial plots of institutions. As in the US, business took an interest, seeing seeds as offering commercial opportunities and profit. Seeds became a commodity.

With Mendel's insights finally embraced by plant scientists, the UK's place in advancing research and development into new varieties of many crops that offered improved yields and resistance to some diseases took off. The Plant Breeding Institute (PBI) was founded as part of Cambridge University's School of Agriculture in 1912. Its first director, Rowland Biffen (1874–1949), was already a pioneer in applying Mendel's laws of genetics with advances in wheat breeding at the start of the twentieth century.

Developed along similar lines to the USDA, the National Institute of Agricultural Botany (NIAB) was established in Britain in 1919 as a purely scientific research element of the PBI. Its focus was on commercial improvements in new varieties. The institute was privatised by Margaret Thatcher's Conservative government in 1987, the result of an ideology that believed, much as American seed producers had a hundred years earlier, that agricultural research should be funded privately rather than by the taxpayer. At the time, 90 per cent of all wheat under cultivation in the UK had been developed and bred at the PBI. The NIAB was not included in the sale; only the plant breeding elements of PBI, which the multinational consumer goods company Unilever snapped up for £66 million. It sold the business on to the American agricultural specialists Monsanto a decade later for £320 million – a prime example of a government selling its assets at a knockdown price, and one that was very profitable for Unilever and its shareholders. In 2004, Monsanto sold its European cereals business to RAGT, a French company that started life as a cooperative in 1919 and today is working with German pharmaceuticals and biochemicals giant Bayer to develop hybrid wheat. The NIAB was privatised in 1996 and is now a charity, supported by its members, made up of farmers and plant breeders. It remains a globally important independent research organisation in plant science, crop evaluation and agronomy. One of its key roles is as a seed certifier for UK plant breeders.

The Arrival of the F1 Hybrid

Both traditional and molecular plant breeding have always been about improving the character of a variety by focusing on specific traits that matter to the farmer: most obviously vigour, yield, disease resistance and flavour, but some breeders may also want to develop crops of a certain colour or plants that grow to a certain height. To achieve this, scientists select

Breaking the Mould

from offspring with specific traits known as pure lines – the result of inbreeding.*

When scientists first applied the principles of genetics to plant breeding, they discovered that if they created a multitude of different lines of highly inbred varieties with very specific beneficial traits and crossed them with others showing different beneficial traits, they would end up with a cultivar that was greater than the sum of its parts; this is known as hybrid vigour or, more formally, heterosis. Successful crosses that were then commercialised were known as F1 hybrids. However, extensive work growing lines over many generations delivered considerably more disasters than triumphs. Unlike open-pollinated varieties, the offspring of F1 hybrids, known as F2s, did not breed true, meaning you didn't get the same crop the second time around.† So farmers got no benefits from saving their seeds, or so they were told by the seed companies, who of course were interested in selling them fresh seed every year.

The result of this approach to breeding has been to create homogeneity – uniformity – in new cultivars in which every individual is genetically identical to its neighbour. It is this homogeneity, realised by narrowing the genetic diversity of the cultivar, that is fundamental to monoculture. Yet heterogeneity,

---

* Inbreeding occurs when the mother plant is fertilised with pollen from its own flowers. On the face of it, this sounds like a recipe for disaster. We think of inbreeding as being something that narrows genetic diversity, which comes with greater susceptibility and weakness. In fact, many plants are natural inbreeders – tomatoes, wheat and peas being just three.

† As we shall see in chapter 7 on page 144, dehybridising – growing F2s and their offspring using classical breeding methods – is proving a very successful strategy for a new generation of freelance breeders.

which describes a crop that displays genetically determined variability in traits or attributes (such as flowering period, disease resistance and seed size), offers breeders greater potential to build resilience and adaptability into new cultivars.

With the advent of the F1 hybrid, plant breeding became a fully commercialised and profitable business, although most of the research and innovation was continuing in the public domain – anyone could find out about it. In order for seed companies to maximise the value of their new cultivars, they needed to ensure no one else knew what varieties were being used to produce the pure lines that formed the parents of the F1 seed. Seeds became secret. Possession was nine tenths of the law.

The Birth of the Patented Plant

Henry Bosenberg (1883–1962) was a rose grower, and in 1931 he applied to the United States Patent and Trademark Office (USPTO) for a patent on a repeat-blooming rose he found growing in his garden. There was nothing unique about the rose; it was a naturally occurring anomaly found in a popular type of tea rose that had been first bred in the US by Walter Van Fleet (1857–1922). Bosenberg was no plant breeder and merely increased his stock through conventional propagation. He named his rose New Dawn and it is still sold by rose growers today. The USPTO took the view that his rose was a unique anomaly among a relatively stable population. This 'new and useful improvement' of a Van Fleet rose cultivar was the first ever plant to be issued a patent.

Bosenberg's success in gaining his patent fired the starting gun for a messy and very poorly regulated global environment that snatched ownership of germplasm out of the hands of farmers and placed it into those of industry. Originally, these patents in the US were given only for plants that could be vegetatively propagated or cloned; the USPTO considered that plants that relied on pollination were not genetically stable.[2]

This ability to apply patents to seeds and treat them as intellectual property became the start of what many, myself included, consider the great steal. The International Union for the Protection of New Varieties of Plants (UPOV), an intergovernmental organisation based in Geneva, Switzerland, was established in 1961 and has been revised several times since then. It was set up to provide an effective system for protecting plant variety by creating a form of regulation to be implemented by its members according to international law. Today, the UPOV has seventy-nine member countries. One of them is the UK, which passed a law, the Plant Varieties and Seeds Act 1964, to comply with the UPOV regulations. Four criteria were used around which new varieties could be protected: novelty, distinctiveness, uniformity and stability. Today, all new cultivars bred by seed companies have to pass tests and be registered on an approved list to show they are distinct, uniform and stable (DUS); the rationale is that farmers need to be certain that the seeds they are buying will perform according to the claims of the breeder.

Patent law is complex and the details are beyond the scope of this book. It is also not universal and is interpreted and applied in different ways in different countries. The European Patent Convention (EPC), signed in 1973, provided the legal framework for the granting of European patents by the European Patent Organisation (EPO), which came into force in 1977. Today there are thirty-nine members, but the organisation is entirely separate from the EU and, at the time of writing, includes agreements with Morocco

The Accidental Seed Heroes

and Tunisia. One of its jobs is to grant patents on seeds and this is causing fury in the world of traditional plant breeders. Although the UK is no longer a member of the EU, it is a member of the EPO and to that end the rules in the UK are the same as across the Channel.

In the US, patent law is interpreted and applied very differently from the UK and the EU, and, as we shall see, is having hugely restrictive effects on independent plant breeders' attempts to innovate. Working with germplasm that was once considered a public good and freely available for them to develop new varieties can end up putting them in the dock, sued by a global seed behemoth with very deep pockets.

Open-source and freelance breeding have been hugely innovative, leading to new generations of crops that are superbly adapted to local growing conditions. Highly experimental breeding using the latest in genomic techniques is undertaken by publicly funded bodies like the John Innes Centre (JIC) in the UK and universities and independent research bodies like the NIAB, not only in the UK but all over the world. These entities, whose gene banks contain thousands of varieties of every type of crop imaginable, make all their work available to anyone who wants to take it on for further improvements and to use it to develop new strains of crops. This resource is the result of decades of collecting and sharing between great institutions and gene banks around the world. Today, open-source research led by the public sector and funded by taxpayers is being given away to the private sector with no strings attached.* Commercial seed companies continue to innovate and improve those

---

* There are some notable exceptions with institutions licensing their creations, for example with apples, as we shall see in A Battle for Domination on page 214 in chapter 11.

Breaking the Mould

genetic resources, which is fundamental to their business model. However, they are under no obligation to share their data and learnings with the institution from which they received the unique seeds. A few tweaks here and there, much work in bulking up, trialling and marketing, as well as having the new cultivars certified and patented, allow the company to claim absolute rights over something that could only ever have come into existence thanks to public money. Ensuring a diverse and resilient future for breeding in all its forms is under attack.

The Problem with Patented Seeds

The rules applied to the patenting of plants and animals are clear: breeding that employs essential biological processes of conventional breeding cannot be patented. (The only exemption is for genetically modified or engineered plants.[3]) Conventional breeding is a process with outcomes that cannot be directly controlled by technical means, including crossing and selecting as well as random mutagenesis (the use of chemicals and radiation to trigger mutations). So, breeding that does not introduce a new trait from a different genus of plant, animal or other organism cannot be considered a technical intervention within the meaning of patent law; therefore, it cannot be patented. But that hasn't stopped agribusinesses from applying for patents for conventionally bred varieties – and getting them. Although the rule is clear, the interpretations can be muddied, and it is down to individual members to clarify. The battle lines are being drawn between pressure groups and the EPC, which does not appear to be following its own directives. Seed patents pose a real and present danger to plant breeders. The application of patents on genetic sequences that are already present in

The Accidental Seed Heroes

traditionally bred crops is becoming an ever greater threat to the ability of small breeders to innovate.

For example, KWS, a German plant breeder and the sixth largest in the world, has obtained a patent for a cold-tolerant maize cultivar, allegedly achieved using marker-assisted selection, whose primary use is as an animal feed with improved digestibility. The patent includes a specific DNA sequence, involving a single gene that confers cold resistance. KWS claims its new cultivar is patentable despite the fact that it is the result of randomly mutated genes. Naturally occurring genetic variations were used in the breeding programmes, which the company claims to have invented.

KWS's claim has repercussions for companies such as the Dutch firm Nordic Maize Breeding, which focuses on breeding cold-tolerant varieties for low-input organic farming and produces about a third of all organic maize seed sold in the Netherlands. Co-founder and director Grietje Raaphorst-Travaille makes the point that varieties of maize with the traits claimed in the KWS patent have existed for years, so should not be subject to patent law. Resistance to a stress like cool temperatures is multifactorial – it involves many genes – and thus the patent goes against the common understanding that traits are expressed through several genes or sequences of genes. Grietje's fear is that defending her varieties against a charge of breaching KWS's patent could prove too costly and she would have to abandon her work, marking the end of traditional maize breeding not only in her country but around the world.[4]

Allowing patents of non-transgenic seeds impacts all breeders, including someone like me who is tinkering with creating new varieties in the back garden. This stifling of innovation acts directly against the need for greater diversity of crops and global food security and represents one of the

Breaking the Mould

greatest impediments to building a sustainable future for plant breeding and food production. It has to be stopped.

In the last half century, this paradigm shift in who controls seeds has only intensified – and spread across the globe. The intellectual property (IP) rights that seed companies claim are limited to twenty years. After that anyone can reproduce the seeds or use them as a genetic resource for further breeding and research so long as the new varieties they develop are distinct, uniform and stable. Plant breeders have been able to avoid releasing their IP, however, by claiming 'utility patent protection', which prohibits other breeders from using their seeds for further research. I'll go into this in more detail in chapter 9, but in a nutshell, it means that after the twenty years are up, all a breeder has to do is make a minor change to a variety for it to be considered entirely new and subject to a new patent. This locks other breeders and researchers out of access to germplasm that could be of real benefit to a more equitable and sustainable future of farming, especially at a local level, and the grower is trapped in an endless cycle of having to buy the latest new cultivar.[5]

I entirely understand why we need a system ensuring limited use of plant breeder rights to incentivise innovation. After all, plant breeders have to make a profit. This is something that taxes many in the world of open-source and participatory plant breeding as well as those, for example, who are creating new populations of heterogeneous cereal crops. The important point is that plant breeders' rights (PBR) should not prevent others from developing new and improved varieties using these seeds. Under PBR, a seed company cannot sell or reproduce the cultivar that is protected without paying a licence fee that, like a patent, is of limited duration, which to me seems entirely reasonable.

The Accidental Seed Heroes

Where the Power Lies

So, who controls what we grow? Michael Fakhri is advisor to the UN as its special rapporteur on the right to food. The reality today, he says, is that seed production has undergone a seismic consolidation. Today, just five seed companies – BASF, Bayer, Corteva Agriscience, Limagrain and Syngenta Group – control over 40 per cent of seeds sold around the world.

Leaving seed production in the hands of large corporations makes transparency very difficult. By design, they are hard to hold accountable. These organisations are in the business of breeding crops that reduce biodiversity, exacerbating global decline. In fact, they are causing even greater harm by dominating what seeds are available. When such a small number of enterprises are gatekeepers to the majority of seed, they also control the price. The world has few, if any, ways to ensure a fair market or even that seeds can be shared freely as they were until the end of the nineteenth century. Current evidence suggests that this domination of the market has not yet seen any obvious increases in the price of seeds to farmers, whereas it is clear that when more farmers share seed or buy from publicly owned plant breeders there is downward price competition.[6]

Whose Side Is the Law On?

Farmers have been put under ever greater pressure by the commodification of seed, being told they cannot save it, share it or give it away: they must buy fresh every year. The world needs to change the political and economic environment in which seed production currently exists. Companies rely on the innovation and work of farmers in the field and publicly funded institutions to supply them with the genetic resources they need to develop cultivars, which they then claim as their

Breaking the Mould

intellectual property. As a result, farmers, who have been the custodians of genetically diverse seeds they have improved and adapted themselves over generations and which continue to evolve thanks to their efforts, are denied a share in the benefits. Michael Fakhri insists that farmers' rights should be prioritised over those of the profit-driven multinational corporations. Indeed, he goes so far as to say that we need to terminate the international intellectual property regime because it does not serve the interests of farmers or the public. All it does is secure greater profits for the largest commercial plant breeders and perpetuate a system of agriculture that is ecologically unsustainable.

According to Bayer, there are 7,000 seed companies serving farmers worldwide, so they argue there is already plenty of diversity in plant breeding. But Bayer alone controls over 17 per cent of the global market. So, 6,995 of those companies have just over half of the market – an average of 0.008 per cent each. Bayer says it empowers farmers to produce more and better to ensure food security and also to offer varieties that require less of Bayer's chemical crop protection. This runs counter to Fakhri's view. On the one hand, you have agribusinesses saying that farmers love what they are doing and it is all about making the planet a better place, whereas what Fakhri and others, myself included, see is overproduction of a limited range of new, genetically narrow cultivars that tie a farmer to a vertically integrated system of cultivation in which the seed company provides the seeds – the least profitable part of the business – and all the chemical inputs, which are the most profitable element. This is a business model designed to perpetuate the status quo and create a dependency culture for the farmer, who becomes locked into just one way of growing crops.

27

The Accidental Seed Heroes

The agribusinesses' argument continues: developing new seed varieties takes up to ten years and costs billions; patent protection lasts only twenty years, so they have just ten years to recoup their investment and make a profit. They maintain that farmers want their – the agribusinesses' – product because they get a better return on every level. For modern, highly mechanised farms in the Global North this may be for the most part true, although there is the law of diminishing returns, which traps intensive arable farmers into using ever more inputs to deliver yields that keep them barely solvent. It is definitely not the case across the rest of the world, where farmers are battling, often against their own governments, for the right to retain control of their means of production, both of seeds and of how they grow food. Evidence suggests that farmer-led breeding and improvements deliver far greater food security and at little or no cost to the farmers, as well as far more innovation in breeding for local conditions. That is not a business model that works for giant agribusiness, but the world needs it!

Big Things Come in Little Packages

The world still gets most of its food from peasant farmers and smallholders, though there is an ongoing debate about the percentages and how the data to determine these numbers has been reached. A report published in 2014 estimated that small-scale and family farmers – those whose holdings are 5 hectares or less – make up 95 per cent of all farms worldwide on 20 per cent of the farmland.[7] In 2009, the ETC Group, an international non-profit that researches on the realities of the industrial food chain and small-scale producers, calculated that 50 per cent of all the world's cultivated food (as opposed to simply a measurement of calories) that goes directly into feeding us is grown on farms of 5 hectares or less. It does not include crops that are grown as animal feed, biofuels or for non-food purposes. These figures became the basis for setting global food policy with the United Nations Food and Agriculture Organization (FAO).[8] But then two papers published in 2018 and 2021 caused much controversy and sought to debunk the numbers.[9] Average farm sizes vary hugely from one country or region to another, but their research suggests that around the world smallholder farms, defined by them as being up to 2 hectares (4.5 acres), produce between 28 per cent and 31 per cent of all crops and a third of the global food supply measured in calories.[10] All the data suggests that small farms have a greater cropping intensity, as seen in higher yields, than larger farms. Furthermore, 70 per cent of calories produced on the smallest farms are directly consumed by humans. The larger a farm becomes, the more of its output (up to a third in the case of the world's largest) goes into animal feed and processing.[11] Revising the percentages downward has reinforced a narrative propagated by agribusiness that the solutions to the challenges of feeding the world should be led by their business model and that small farms are neither as efficient nor as productive as large ones: they claim small-scale farming is inefficient and cannot feed the world. Our future, so their argument goes, depends on industrialised agriculture using the seeds that are bred to support this system. In 2021, the FAO estimated that 35 per cent of all our food is grown on farms of less than 2 hectares, utilising just 12 per cent of farmland globally. However, the methodology used to get to these numbers warrants considerable further scrutiny.[12]

Family farmers and smallholders are embedded in their communities, as I have seen in my travels around the world.

The Accidental Seed Heroes

Crucially, smaller farmers are also custodians of their country's genetic resource: seeds they are continuously improving and conserving. But this cohort has the least bargaining power when it comes to standing up to, let alone negotiating with, their governments to put their interests first. These farmers are not only the ones who feed themselves and their communities but are protectors of biodiversity because of the way they produce food. Giant farms that are found not only in the Global North but across much of South America and southern Africa are dependent on growing the latest cultivars for the most part as vast monocultures that underpin a business model of modern plant breeding, which I am not alone in believing is unsustainable and fundamentally bad for our planet.

Many commercial plant breeders see the need to work with wild relatives, FVs and traditional crops in developing new, climate-resilient varieties. They need to be mindful of their sources of material and recognise that their most important stakeholders are not the company's shareholders but the farmers who have the societal connections with this vital gene pool. It is ever more essential that plant breeders large and small – and that includes accidental ones like me – remind ourselves that the intellectual property of those native, local and indigenous foods rests with the people who have maintained and improved them through countless generations of diligent selection. They have to share in the benefits that others receive as a result of working with their seeds. Germplasm held in publicly funded institutions can be accessed for free by anyone wanting to develop new varieties. The giant agribusinesses, with often minor tweaks to the genetics, can claim them as new inventions over which they control the IP of the entire genome: a perversion that has to stop.[13]

CHAPTER TWO

Where Farmers' Varieties Reign Supreme

On the Trail of Deliciousness in Albania and Ethiopia

We used to grow the best oranges in Ethiopia and the coffee that grew alongside had them in its aroma. When the oranges would grow no more, we planted avocado instead and now the coffee has them in its aroma.

Ducamo, a farmer from Hanfa village
in Sidama, Ethiopia

There were two countries I was particularly keen to visit when researching for this book – Albania and Ethiopia. They both offer insights and act as exemplars for us all to see how we can build sustainable and nature-friendly ways to feed ourselves. What I was to learn from travelling to them stemmed from the genius and persistence of their farmers.

'What are you going to discover there?' That was the question put to me by a curious friend who wondered why, of all the countries on the African continent, I had chosen to explore a small part of Ethiopia. Simple really. The world needs more climate-resilient crops, most especially in countries like Ethiopia, which are most at risk from climate change. It's a special part of the world, too. Some of Africa's greatest rivers run through it; its landscape ranges from hyper-arid desert to tropical and subtropical arable land rising to 3,000 metres; it is home to the greatest diversity of indigenous and native domesticated crops in all Africa and some of the most exciting, which demonstrate agricultural strategies that not only feed a country of 120 million but also have the potential to feed much of the world. I wanted to remind myself that many of the solutions to achieving net-zero carbon future for food lie in traditional plant breeding that champions crop diversity and its unbroken link to the dawn of agriculture.

An African Exemplar

Where I was travelling, through the southern part of the Ethiopian Rift Valley, at an elevation between 1,400 and 2,800 metres, the soil is fertile and the weather, in March, redolent of a blissful English summer. Ethiopia is home to one of Africa's oldest and largest seed banks and works with at least thirty community seed banks across the country, run by farmers for farmers to maintain and share FVs of most crops. In the country's informal seed market, farmers are the ones who hold the key to a sustainable and secure food supply. At the Ethiopian Biodiversity Institute (EBI), Dr Tamene Yohannes buys and acquires seeds of adapted and local varieties, which the institute shares with community seed banks. Farmers then borrow these seeds and at the end of the harvest return the amount they borrowed

Where Farmers' Varieties Reign Supreme

plus a little extra as interest. At no time does money change hands. The farmers only return seeds that meet the bank's quality threshold, ensuring stability and consistency in these varieties.

The FVs in circulation result from generations of conservation and selection – a practice that is replicated by countless millions of indigenous farmers all around the world. In Ethiopia, this is a dynamic and continuously evolving process of genetic conservation and improvement, which is constantly responding to the effects of climate change; all at no financial cost to the farmer, who is in full control of the means of production. Conservation of these priceless FVs works because the seed banks are at the core of maintaining diversity, with seeds being shared every year within the community. In an informal agroecological process known as 'conservation through use', crops are adapting as they go, resulting in a great diversity of these FVs. In fact, 90 per cent of agriculture in Ethiopia employ them.

Ethiopia is not alone in operating community seed banks. Similar things are happening at scale in India; would that more of the world embraced a system of seed production that supports biodiversity, soil health, resilience and farm income, and which, above all, can feed us all.

In Ethiopia, FVs are known as the father and mother of commercial varieties. Plant scientists working at the EBI and in agricultural research departments recognise that modern plant breeding needs to play a major part in the future, with more mouths to feed and ever-greater challenges to cultivation; FVs have a key role. This is especially the case with wheat and maize, because the Ethiopian government sees the export of these crops as a critical plank of economic policy, while also understanding that commercial varieties cannot be developed without access to the diversity of ever-evolving FVs. It is a common theme: the solutions to global food security are dependent upon a diversity of approaches to breeding and improving varieties. Traditional methods, science-led innovation and modern plant breeding should exist side by side. The commodification of seeds as perpetuated by giant seed companies has precisely the opposite effect: homogenising crops increases food insecurity not just for Ethiopians, but for all of us.

The EBI holds germplasm of ninety thousand varieties of edible and medicinal plants, both wild and domesticated. Some seed is multiplied at local research centres and then given out to farmers; many are held in other agronomic institutions across the country, as well as in the Global Seed Vault, a vast collection buried in a mountain on Svalbard, an island above the Arctic Circle, whose purpose is to guard the world's food supply against catastrophic loss. Ethiopian farmers are determined to maintain their seeds and there is a central seed restoration programme to protect against varietal loss, known as genetic erosion, which may occur if farmers are displaced because of wars and famine.

Preserving What Is Left

Often friends and family fear that my seed-detecting adventures are ill advised. 'It's dangerous and full of criminals and people trying to get to England to escape,' was a response repeated many times by people I told about my plans to visit one of the best-kept agricultural secrets in Europe – Albania. Like Ethiopia, it is a country most of us know little or nothing about but are happy to hold

Where Farmers' Varieties Reign Supreme

prejudices against. It, too, is home to a great diversity of crops and flora because it is the last of the Balkan states to have emerged from the yoke of Communism and introduce modern hybrid cultivars to replace traditional and locally bred ones. Since 1995, the government has made no investment in plant breeding, focusing only on importing seeds of new cultivars. Yet under the rule of Albania's autocratic dictator Enver Hoxha (1908–1985), when the command economy – in which a government sets, among many other things, permitted levels of production, terms of distribution and pricing – determined what you would study, agronomy and animal husbandry were considered the most important subjects. Only the best and brightest students were selected for undergraduate study that included all aspects of food production and plant breeding. I was eager to meet one of that cohort, Professor Sokrat Jani, now director of Albania's own gene bank.

Tirana, the country's capital, can be a noisy and somewhat chaotic city to get around, but once the car I was in had fought its way through the traffic, the quiet courtyard of the Albanian Institute of Plant Genetic Resources, part of the Agricultural University of Tirana, was not just a blessed relief, but also a portal into an orderly world of seed saving: Professor Jani's domain. His gene bank is home to five thousand accessions (collections of plant material from a single species) of cultivated plants in Albania. But as he was keen to point out to me, there are thousands more FVs in situ – in other words, being grown and maintained, but in ever-decreasing numbers, by the remaining traditional farmers cultivating their crops in some of the remotest villages in the country. In every community there are local variants of most vegetables, and time is running out to save these undocumented gems. And save them we must. They are the bedrock of local food culture and fundamental to a traditional form of farming that is a way of life for the thousands in the country who work their small plots. Most importantly, these FVs are the genetic resource we all need to develop and improve, varieties with adaptations that have evolved over generations, a vital resource to reinforce our ability to grow food as the climate becomes ever more extreme.

It's not only the farmers who have been saving FVs, passing their skills from generation to generation, who need support; Albania's institutions need help too. The resources available to Professor Jani and his team are limited, which is why he holds only a fraction of the varieties from Europe's last stronghold of genetic diversity of locally adapted crops. He needs a new generation of enthusiastic and committed agronomists and botanists to pick up the baton and travel into the numerous isolated communities to find those lost relatives of the fruits and vegetables that were grown in abundance as little as fifty years ago. Professor Jani has discovered many forgotten and abandoned FVs himself, including tomatoes that have reverted to the wild and been self-seeding in eastern Albania for one and a half centuries. One can but imagine what traits they might have that could prove invaluable to breeders.

Professor Jani worries the most about what will happen when returning Albanians bring what he calls alien seeds back home and into his country's beautiful mountainous regions. These imported varieties can and do end up replacing the traditional FVs farmers used to cultivate. I have encountered this sadly familiar story repeatedly in my travels around the world as a seed detective. Switching

Where Farmers' Varieties Reign Supreme

to these imports further accelerates the demise of locally adapted varieties.

This calls for action on two fronts. Firstly, Albania needs to put more resources into collecting and describing local varieties and securing their future by both growing in situ and storing ex situ in local gene banks. Already the Global Seed Vault on Svalbard holds some four hundred wheat, four hundred maize and two hundred common bean (*Phaseolus vulgaris*) accessions from Professor Jani's collection. Secondly, the country needs to support small-scale farmers who are custodians of local varieties to improve diversity and resilience, especially in those places where so-called alien invaders might add unwanted genes. Restoring viability in old seeds is also best done by growing them in the places they come from. And, of course, Albanians and visitors alike should be eating and enjoying them.

Albania also urgently needs to encourage a new generation of farmers to embrace the unique place of the country's food crops as exemplars of local adaptiveness, to build resilience both botanically and economically. Farmers who select and save the seeds of their region's FVs are true seed heroes and represent the very best in maintaining and improving these FVs. I was to meet many on my travels.

Professor Jani told me that, after forty-five years of research and conservation, his passion for vegetables is not dimmed. His real concern is to know where the next generation of agronomists and geneticists will come from. With dwindling interest among undergraduates, who will continue the work he has been leading for so long? The need to ensure the diversity of locally adapted FVs and cultivars

37

that were part of an active breeding programme until the mid-1990s is greater than ever. Enver Hoxha succeeded in making the country self-sufficient in food because he ruled with an iron fist for over forty years. Since the collapse of the Communist regime in 1991, Professor Jani's department has been starved of money and talent. Fortunately, the World Bank gave a grant to his university to ensure the survival of the gene bank and now he has some twenty-year-old deep freezers and conservation space for the country's collection of edible crops.

Banking on a Seedy Future

I was keen to see an Ethiopian community seed bank in operation. Dr Basazen Fantahun is a wheat breeder at the EBI who is also working with triticale – a hybrid cross between durum wheat and rye. He is developing varieties that display wide adaptability for highland areas. After three years of testing in fifteen different agroecological environments, his first elite variety, Simit, is proving popular. * Seed is supplied to local farmers through the seed cooperative. They save seed for the first three or four harvests and then purchase fresh to ensure varietal stability (it is easy for seeds from other varieties grown nearby to get mixed in accidentally during harvest and cleaning. Disease resistance also deteriorates in seed that is not well maintained).

Dr Basazen suggested I visit a community seed bank at Chefe Donsa, a small town a two-hour drive east of Addis

---

* An elite variety is one that has been selectively bred using modern genotype techniques to exhibit traits considered superior to traditional varieties.

Where Farmers' Varieties Reign Supreme

Ababa. I wasn't to be disappointed. My driver, Ermias, and I arrived late morning. The small compound, behind a solid metal gate off the dusty main street of the town, consisted of a converted cattle shed to store the harvest, a small office and a cool room to test and keep examples of the crops being shared – all in carefully labelled airtight jars. The seed bank is administered by the farmers' co-op, which has 1,500 members, who altogether grow on 6,000 hectares. Surplus seeds are sold to non-members to help fund the operation. As we arrived, a farmer was in the process of offloading three 50-kilogramme bags of FV wheat from his long-suffering donkeys, who, relieved of their loads, scuttled off to graze while he had the weight checked and samples taken for analysis. Other farmers were gossiping in the grain store, which was also home to chickpeas, red lentils, common beans, barley, maize, herbs including caraway, coriander, and green and white fenugreek, and medicinal plants too. The shed was filled with mountains of sacks, each containing different FVs. I spent a convivial half hour being shown their harvest by the farmers who also administer the seed bank. They delighted in pointing out the most delicious varieties. These included black chickpeas – everyone's favourite when used to make sweet cakes; a local FV yellow chickpea called Shimbera; a white one called Aborti; and one used for fattening cattle called Guaya. I was passed handfuls of seven different varieties of brown wheat known as *geja*, each having a particular culinary use and selected by women, who are the cooks in the family. I admired a variety of black barley known locally as Keselee – excellent roasted, I was told, and just one of a hundred barley varieties grown by the members of the community. The local bread wheat, called Kovrit, is preferred baked as a whole grain. Some in the shed were growing another brown

The Accidental Seed Heroes

FV wheat, Kukura, and an improved cultivar, Mangudu, that they had visions of exporting. As we shall see, saving and maintaining FVs is a way of life for some in Albania too.

Among the Seed Savers

It was a scene I had only ever dreamed of witnessing: an earthy quilted valley, no more than a quarter of a mile wide, filled with a mosaic of vegetables, surrounded by gently rolling hills of small wheat fields ablaze with poppies and cornflowers; the verges dotted with wild acacia in bloom – a tempting snack – and everywhere wildflowers and the heady perfume of thyme and oregano. I was just outside the village of Miras, deep in the Korçë County of southeast Albania.

My idea of heaven was the 250 acres (100 hectares) of cultivated ground that provide a livelihood for the fifty-plus families who work the land in Miras and who produce more than enough food to feed the entire community of six hundred families. I was there with Lavdosh Ferruni, a passionate advocate of sustainable, organic agriculture in Albania, and his associate, Professor Robert Damo, an agronomist and botanist who heads the Faculty of Agriculture at the Fan S. Noli University in Korçë. Like Professor Jani, Professor Damo was an elite student in the 1970s, who found his calling working in the most important sector of the economy. He has spent his career researching, cataloguing, trialling and maintaining the great diversity of FVs that continue to be grown throughout the area: a centre for locally adapted, genetically and organoleptically* unique FVs in the country.[1]

---

* Organoleptic describes the qualities of a food that stimulate the senses, including taste, colour, smell, appearance and feel.

Where Farmers' Varieties Reign Supreme

The fields of Miras have been continuously farmed since Roman times, and there is archaeological evidence of agriculture dating back 4,500 years. There is a very long tradition of saving seed and selecting for favourable traits throughout Albania, but the Korçë district is the last stronghold. The loss of locally adapted FVs with enduring historical and cultural significance varies from village to village. In some cases, at least half the genetic diversity has been lost in the last thirty years; in other localities it is as much as 75 to 80 per cent in the same timeframe.

Yet unique and important varieties survive. Robert was keen to show me the FVs he had been collecting over many years from farms and villages in the district. The Albanians like their popcorn, and two wonderful FVs that can be grown without irrigation are the much-prized Korçë black popcorn and Moscow red popcorn, also known as hooked red popcorn. A sweetcorn that also grows without irrigation is named after the lake to the north of Korçë, Prespë. Bean FVs continue to flourish, as do capsicums, tomatoes, leeks, onions, squash, lettuce and orache (*Atriplex hortensis*), a member of the amaranth family, which is similar to spinach and grown as a summer crop; also, white cabbage, which is less common, although I was to come across some fine examples on another visit later in the year. Professor Damo was like a kid in a sweetshop as he strode among the orderly rows of seedlings being planted out by a cohort of elderly but sprightly farmers, asking questions about what was being grown and how crops were being affected by the unseasonal weather.

Albania is also home to an enormous diversity of local varieties of apples, pears, stone fruit and grapes – it has the oldest tradition of viticulture in Europe – and I can attest to

The Accidental Seed Heroes

the wines being a revelation. As indeed was a particular take on pear juice, the result of steeping a local variety with the addition of wild horseradish leaves in water for six months. The flavour was unimpeachably delicious and unique, and the pear tasted amazing too. I was offered as many versions of raki as there were farmers distilling it from the fruits and berries that filled their orchards and hedgerows, almost all utterly delicious.

The Tree Against Hunger

At an elevation of over 1,800 metres, the countryside of Ethiopia's Sidama region is green and glorious: steep, narrow paths meander through a landscape of agroforestry. Coffee trees – indigenous to the country – are everywhere, growing beneath the shade of a New World introduction: avocados. The land is filled with randomly sown FVs of beans and maize. In clearings, the Sidama people have built their round mud houses, thatched with sorghum or wheat straw. Cattle and goats keep the grass short and old men doze beneath the shade of trees planted over the graves of their ancestors. One such is the elephant tree (*Boswellia papyrifera*), so named because its bark looks like the animal's skin and has great cultural relevance. It is a place for conflict resolution between arguing family members because, beneath the tree's canopy, one must speak only the truth. It's a conduit to talk to one's ancestors for advice and also to ask for rain. The tree has another use: it produces a much sought-after lemon-scented resin – frankincense. But there is another tree – the tree of life – that the inhabitants depend upon to sustain them.

One of Ethiopia's most important indigenous crops is enset (*Ensete ventricosum*). Known locally as the false

Where Farmers' Varieties Reign Supreme

banana, it is a member of the same family, Musaceae. Wild relatives can be found from South Africa, along the eastern edge of the Great African Plateau and as far north as Ethiopia – the only country to have domesticated it.[2] I saw enset being grown across great swathes of hillsides that I travelled through. Enset is a very important staple for the twenty million people who depend on it for their carbohydrate. Its stem, known as *shefina*, and its root, called *amicho*, are fermented and ground up to make porridge, bread and delicious pancakes flavoured with rock salt dug up from one of the hottest and lowest points on earth, Lake Afrera, far to the northeast.[3] I had it harvested and prepared for me to eat by Shanu Libabo, a sprightly grandmother and remarkable farmer living on the edge of the village of Hanfa, in the heart of the Sidama region, a long day's drive south from Addis Ababa.

I was privileged to meet Shanu, who took me into a grove of enset growing behind her home. Being an elder didn't mean she couldn't do the work of someone a fraction of her age. As lithe and slender as a ballet dancer, she set to cutting down the huge, fleshy leaves of a 3-metre (10 feet) high plant, slicing it across the top at about 1.2 metres (4 feet) above the ground. She vigorously chopped around the root and, pushing on the stem, deftly removed the root ball. She lifted its 10 kilogrammes (22 pounds) as if it were a feather and, cross-legged, set to work scraping off the fibres and flesh from the leaves, collecting the water that poured out of them. With her razor-sharp small machete, Shanu chopped the root into a pulp, which she pounded and placed in a depression in the ground covered in enset leaves to ferment for a month. I felt exhausted just watching her. She didn't break a sweat, and then she took me inside where she ground and sieved some of the harvest from a previous felling she had brought out of her store. Over a charcoal fire, she then cooked me a yummy pancake, washed down with coffee made with beans from trees in her part of the forest, roasted and ground by one of her granddaughters.

A Remarkable Root

Enset's place in Ethiopia's food economy cannot be overestimated. It is a very important source of carbohydrate and the root is rich in calcium. Some 85 per cent of Shanu's community eat enset porridge – *boula* – every day. The fermented juice is quite alcoholic and is given to the cattle because it helps with milk production! It is also boiled down to make glue. Enset is a perennial and is easy to propagate – the base of the plant is buried upside-down and numerous suckers emerge within a few weeks. It is invaluable, too, because you get so much nutrition from a very small space. It is drought tolerant and, once prepared in the way Shanu showed me, can be stored for more than six months. Because it can be harvested at any time of the year, it is an essential survival food against drought; its leaves and stems are an important fodder crop, while the fibres left behind after processing are used to make rope and as a packing material. In Sidama, research has shown that families who grow enset are far less likely to suffer from malnutrition than those who do not.[4] Altogether, enset stands out as one of the most important subsistence crops of Ethiopia, sustaining over 20 million people – one in six of the country's 120 million population – and one that could play an important role in the future as part of a more biodiverse and secure food system.

Where Farmers' Varieties Reign Supreme

Enset flowers when the plants have reached full maturity – at about ten years old – after which it dies. Farmers normally harvest it once it has reached a suitable size – usually after four to six years – which means that these days few if any specimens get to flower, set seed and thus improve the gene pool. In the past farmers collected seed from favoured specimens, contributing to the diversity of FVs that have emerged over possibly ten thousand years of domestication and selection.

Over half of all domesticated food species – including potatoes and most fruit – are propagated clonally by indigenous farmers and many seed and plant nurseries.[5] This avoids farmers having to select from generations of breeding to stabilise desirable traits. But there is a price to be paid with clonal reproduction: without the recombining and mixing of genes from across a diverse gene pool through sexual reproduction there can be an erosion of the genetic potential that is associated with farmer-led selection. Unwanted traits – disease susceptibility, for example – cannot be eliminated through clonal selection, and adaptiveness to climate change through evolutionary plant breeding, a hallmark of FVs, cannot happen either.[6]

Enset is classified as an orphan crop and as such is of little interest to most plant scientists. A notable exception is found among some ethnobotanists at the Royal Botanic Gardens, Kew, who have been studying it for some years, in collaboration with the University of Addis Ababa. Ethiopian plant scientists and agronomists share a deeper understanding of its social, cultural and nutritional significance. As well as cataloguing and describing the hundreds of FVs and the genetic diversity they contain, they hope to improve cultivars to meet the huge demand for this crop from Ethiopia's growing population.

The Accidental Seed Heroes

Maintaining the diversity of FVs is also challenging. Although enset is a brilliant insurance crop in times of severe drought, during the 1970s and 1980s the amount under cultivation was dramatically reduced because plants were harvested when immature and not replaced in sufficient quantity to meet demand. An indigenous method of agriculture that had evolved over millennia to provide food security in the face of environmental extremes of drought and flood was in trouble. Population pressure has increased and because of it farmers are now growing more enset than ever – up 61 per cent between 1997 and 2019, with the greatest increase being in the most drought-prone regions.[7]

Ethnobotanists have also been learning that Ethiopia's indigenous farmers appear to be responding to climate change by adapting the ways in which they grow their crops – especially underutilised orphans like enset – by changing planting times and selecting from plants within a harvest that are showing improved performance in less reliable seasons, offering insights that are equally relevant for all indigenous agriculture around the world. I came away from my first foray into southern Ethiopia in awe of the farmers I met. They are first and always the people we must listen to and learn from as we who inhabit the Global North adapt the way we grow crops as our local climate changes too.

Albanian Farmers' Varieties Under Attack

The impact of a market economy and depopulation, as a result of a movement to urban centres, emigration and the general disinclination of young people to work in agriculture, has had a devastating effect on FVs in Albania.

Where Farmers' Varieties Reign Supreme

Until 1990, the system of collective farms in a country that had been able to utilise its abundant water to irrigate most farmland meant Albania was able to feed itself, albeit poorly. Collective farms grew local varieties and saved the seed that flourished in the varied growing environments across the country – although there was little, if any, incentive for those running the farms to innovate or maintain the diversity of crops they had under cultivation, which was probably one of the reasons why food was often in short supply. The reallocation of collective farms after the collapse of Communism meant that thousands of families were given land to cultivate themselves. Today, the average holding is 1.4 hectares (less than 2 acres), and about 35 per cent of the working population of around 1.5 million are employed in agriculture – until 2015 Albania's primary economic sector. The country continues to grow nearly everything it needs, and the public prefers to eat locally grown fruit and vegetables.

As in so many other countries I visited, where agriculture was built on a system of small farms growing local varieties and saving and sharing seed, changes in practice driven by the introduction of modern hybrid cultivars have resulted in the collapse of varietal diversity. This has happened because once a farmer stops growing and saving seeds of local varieties and grows only modern cultivars, the traditional ones are abandoned, disposed of or simply fed to the animals. In just one season, unique FVs can be lost for ever. Albania is suffering this fate.

In the last thirty years, Albania has moved from growing exclusively local varieties of arable crops, specifically cereals and maize, to importing 80 per cent of its seed. The great diversity of open-pollinated varieties of maize is now lost

– except in a handful of mountainous regions, as I discovered. Outside the major cities, fresh fruit and vegetables still make up a significant proportion of the diet, and local varieties retain a foothold, although their future is uncertain. In 1990, twenty-three vegetable crops, all local varieties and almost exclusively FVs, were grown in the low and coastal regions. Today this has increased to thirty-eight, but the majority of these are hybrid cultivars.[8] The reason for this switch is understandable – modern cultivars offer greater yields, which can justify increased production costs of fertiliser and chemical inputs. Sadly, growers and distributors are paying less attention to the things that matter to consumers: provenance and taste. Albania is in danger of making the same mistakes as many other countries by neglecting, abandoning even, its indigenous food culture.[9] Involving farmers in a process of continuous adaptation and improvement in local varieties by maintaining and selecting for desired traits is not only a country's best defence against the impact of climate change on crop yield, but also the best way of conserving a genetic resource that can prove invaluable for building resilience and diversity both locally and in other parts of the world.

Seeding a Collaborative Future

A decline in the diversity of FVs will, over the medium to long term, have a hugely negative impact on localised self-sufficient food systems. I believe there is an urgent need within many countries like Albania and Ethiopia to focus on participatory breeding with farmers to improve seed quality and strengthen traits that can flourish in our changing climate. More academic and cultural links between well-resourced and globally connected educational

Where Farmers' Varieties Reign Supreme

and agricultural research institutions, collaborating with local post-graduate researchers, would be hugely beneficial, not only to Albania and Ethiopia but to all those who come to these institutions to work and learn.

When FVs are held in university collections, they are available for researchers and breeders to see if their unique traits – abilities to grow in extreme drought, for example – can be used in both traditional and molecular breeding. Other unique characteristics – showing resistance to certain pathogens and diseases, as well as better colour, vigour and deliciousness – are essential tools for plant breeders to develop new varieties and improve existing ones because they have access to a more diverse gene pool than if they only work with existing cultivars. The alternative – depending on genetically narrow and supposedly 'one size fits all' hybrid cultivars bred and controlled by agribusiness to feed us – doesn't, in my view, bear thinking about.

Despite the decline in traditional, sustainable and agro-ecological farming across Albania, I see signs of new shoots. Tourism will be a key economic driver and agritourism a vital tool in restoring economic fortunes in the rural areas, encouraging a new generation of more entrepreneurial young farmers. Celebrating Albania's remarkable food culture, and promoting and expanding its deep history of herbal medicine and its incredible diversity of flora – Albania is home to 4,000 currently identified flora, 1,500 of which are found in the Korçë district alone – will have a transformative effect. Albanians need only to look at the damaging impact of industrial farming around the world for affirmation of why their own food and farming system needs to be nourished. Ethiopia has been a popular tourist destination for the richness of its wildlife, archaeology and

The Accidental Seed Heroes

cultural heritage. Agritourism has huge potential there too, but ethnic and political conflict has discouraged visits from all but the most determined travellers. I can but hope that peace will be restored in that amazing country soon, because its desperately underutilised human and agricultural capital has the potential to transform food production for hundreds of millions of Africans.

CHAPTER THREE

Using One's Loaf
The Bigger the Population, the Better

The greatest service which can be rendered any
country is to add an useful plant to it's culture;
especially a bread grain.
Thomas Jefferson, 'Summary of Public
Service' (post 2 September 1800)

When that loaf comes out of the oven and fills the house with the addictive fragrance of freshly baked bread, it is – almost – the end of a journey that wheat and all its many ancestors have been on for the last twelve thousand years. It is a crop I have never grown but its seeds are something that I consume on a daily basis and where they come from and how they were bred and improved really matter to me. In the world of molecular cereal breeding, homogeneous 'elite' cultivars dominate, but there is another way. Evolutionary and population cereal breeding is on the front line of a counter-revolution that challenges the narrative of the major breeders. It offers a meaningful alternative to intensive monocultures if the future for diversity, resilience, nutrition and the environment, growing one of the world's most important food sources, is to flourish. It's an exciting and challenging

51

The Accidental Seed Heroes

time for the many traditional farmers and breeders who are joining forces to counter the hegemony of agribusiness.

I have been baking bread since I learned from my mother as a kid. As a student at a Rudolf Steiner school I was subjected to slabs of solid, worthy wholemeal – it was that or go hungry – so I was determined to attempt an improvement by baking myself. I cannot say in all honesty that I was any good. Baking requires a Zen-like approach and it was only after studying with the French baker Richard Bertinet, who taught me the art of stretching and folding and the importance of getting the correct hydration (proportion of water to flour), that I finally, some twenty years ago, began to produce passable loaves. As my baking abilities improved, I began to care about the provenance of the wheat I used. For years I thought it had to be French to produce the best bread, otherwise Canadian, but certainly not home grown. British flour was only good for animal feed, or so I had been led to believe. How wrong I was.

A Welsh Stalwart

Despite diligent purchases from organic suppliers, it was only when I was given a bag of a Welsh FV called Hen Gymro that I finally understood that all flours are not equal and that it was a myth – propagated by nearly two centuries of relying on cheap wheat from Russia and the British colonies – that we could not grow decent bread wheat in the UK. Hen Gymro may be a traditional FV, but it is also one of a large number of cereals used in baking – including rye, barley, spelt and emmer wheat – that are being restored and improved using breeding and selecting techniques that have changed little for millennia. For me, baking with Hen Gymro marked

52

the start of a journey discovering how to grow varieties that are the antithesis of modern hybrid cultivars.

Anne Parry has a busy way about her. As a miller she has to be an expert in multi-tasking, but she is also a baker who knows her wheats. She runs the Felin Ganol water mill in West Wales. When I have visited her, the quiet churning of the wheel and the rattle of belts and millstones grinding have provided a minimalist, calming soundscape. Seeing 'my wheat' going through the system, being turned from grain into a rich, aromatic wholemeal flour, I felt truly connected with what I bake and eat. To me at least, Hen Gymro is unimpeachably tasty, wonderful to bake with and full of nutrients essential for a healthy life. It was a revelation when I first used it; by blending it with sifted white flour, I could make a loaf that was a match for anything I had tasted before – even those scrumptious baguettes with their seductive aroma that beckon the hungry traveller passing any boulangerie in France. I have used Hen Gymro for sourdough loaves but my favourite way to bake is with a poolish, when I mix equal quantities of flour and water with some yeast and leave it overnight or until the following evening before baking bread I can eat for breakfast. So, what is this, my favourite wheat, and how did it find its way to Anne Parry and her water mill?

Hen Gymro translates as 'Old Welsh'. It was being grown as late as the 1920s in southwest Wales and is the longest known surviving British wheat FV; one locally adapted version of a diverse population of a variety that had evolved over many generations. In the 1920s, T.J. Jenkin (1885–1965), a scientist at the Welsh Plant Breeding Station at the University of Aberystwyth (and later its director), started collecting the many examples of this FV, discovering over 250 lines, which he planned to use to develop an 'improved' variety.

He and his successor as director, Evan Thomas Jones, noted a number of reasons why Hen Gymro had hung on in the face of fierce competition from the new cultivars being bred at the start of the twentieth century. Firstly, it was principally used for home baking on farms and in small communities that were largely self-sufficient. Secondly, it seemed to get by in the most unfavourable of conditions for growing wheat. Despite a wet and unreliable climate, the various lines consistently produced millable grain, especially when ripening conditions were particularly poor. Finally, the long and robust straw was perfect for thatching.[1] Jenkin and Jones recognised the obvious genetic diversity across the FV, which is almost certainly the reason for its resilience and usefulness, and identified two specific lines, S70 and S72, as being superior. We shall meet these again later in the chapter.

Wheat and Us

Before continuing the story of the place of FV wheats in my own life, reminding ourselves of how wheat became so entwined in the story of Western civilisation and then of world food sets the scene for what is to follow. Wheat comes in many guises, but in all its forms it started its evolution in a region of the eastern Mediterranean and Middle East known as the Fertile Crescent. It was one of the first crops ever grown by our Neolithic farmer ancestors and the story of how it evolved with the help of humankind and its genetic makeup raises the question: was it us who domesticated wheat or was it wheat that domesticated us?

Wheat is part of a family of grasses of the genus *Triticum*. Different species can naturally hybridise and cross with other wild grasses and have been doing so for aeons. Hunter-gatherer

Using One's Loaf

communities had been harvesting and storing these grasses since *Homo sapiens* moved into the Fertile Crescent at least eighty thousand years ago. It has been suggested that humans living in the region were already using a type of flint sickle to harvest wild grasses twenty-three thousand years ago.[2] Using such instruments a family could, over a few weeks, forage more than enough grains to feed themselves for a year.

The first fields of farmed cereals would have been sown from harvests gathered from populations of a great diversity of wild grasses. Experiments suggest that the process of complete domestication of such crops could have happened within two hundred years and arguably as few as twenty to thirty years without any conscious selection by the farmer.[3] These, the first wheats to be domesticated, would have been primitive ancestors to two closely related 'heritage wheats' that are gaining awareness and popularity today because of their health benefits and wonderful flavours: einkorn and emmer wheat. These primitive wild parents are diploid, meaning they have two sets of either seven or fourteen pairs of chromosomes. Wild emmer wheat, *Triticum dicoccoides*, is thought to be the result of a cross between wild einkorn, *T. urartu*, and wild goatgrass, *Aegilops triuncialis*. Goatgrass, like emmer wheat, is a member of the family Poaceae.[4] The two are genetically very similar, which has enabled them to hybridise readily. This crossing introduced two more sets of seven chromosomes into the emmer wheat genome, which now doubled to give a total of twenty-eight, making it a tetraploid – an organism containing four different sources of chromosomes. As Dr Simon Griffiths, group leader for the Delivering Sustainable Wheat programme at the John Innes Centre, told me, this spontaneous doubling is very rare but incredibly important in the story of the domestication of wheat because the combining of the genomes of these different species resulted in new species that were self-fertile.

Emmer was the first wheat to be fully domesticated and its more complex genome meant that the seed remained within the 'ear' or seed pod when ripe, unlike wild wheats, which shatter, shedding their seed, as soon as they are ripe. This non-shattering trait meant that a crop could be harvested and threshed later; being softer, it was also easier to separate the seed from the chaff. Emmer wheat spread from the Fertile Crescent, arriving on these islands about four thousand years ago.[5] Several different types, the result of deliberate selection and local adaptability, emerged over the subsequent millennia. As late as 1940, one known as rivet wheat was still being grown in England.

The evolution of wheat was happening on several fronts. At the same time as cultivated emmer wheat appeared, another hybridisation occurred between it and goatgrass, giving us durum wheat, *T. durum*, which is used to make, among other things, pasta. But the most significant development was the emergence of bread wheat, *T. aestivum*, the result of hybridisation between the new forms of cultivated emmer wheat with its twenty-eight chromosomes and another wild goatgrass from the Middle East, *A. squarrosa*. This added another two sets of seven chromosomes, resulting in a genome containing forty-two chromosomes: this new wheat is a hexaploidy, evolved from three parents – the equivalent, according to Dr Griffiths, of us having three kidneys!

What makes bread wheat so special is that its much-enlarged genome contains many chain-like protein molecules known as glutenins. Diploid species like goatgrass have just two or three genes that contain these proteins, whereas bread wheat contains all the glutenins from the genomes of its parents. It is the gluten in bread flour that gives it its stretchability, so the dough retains the gases produced by fermentation, resulting in an airy loaf! Another really important trait of bread wheat is that its genome has the ability to adapt to a wide variety of growing conditions. It was born to be continually improved and developed through natural and deliberate selection by humankind. It is at this point that I wonder if it was our ancestors who domesticated wheat or bread wheat that domesticated us. We played no part in the accidental hybridisations, although ethnobotanist Dr Alex McAlvay of the New York Botanical Garden in ongoing research suggests one wheat species, *T. zhukovski*, first identified in the southern Caucasus, may have originated from human-selected ancient wheat populations, demonstrating that bread wheat could continue to evolve and change by our hand.

Wheat's ability to adapt, and the fact that it had become a staple food, ensured that vast populations of FVs would evolve over the millennia. These would show an almost endless number of traits, covering every imaginable colour and shape of grain, length of stalk, taste… What is left of this diversity is a vital resource for developing new cultivars by refreshing the gene pool in molecular breeding and, as we shall see, in building evolutionary populations. The N.I. Vavilov All-Russian Institute of Plant Genetic Resources (VIR) in St Petersburg holds nearly 40,000 wheat accessions alone.[6] The gene bank in Aleppo had a collection of 150,000 accessions of crops from the Fertile Crescent. Fortunately, this collection was brought out of Syria in 2012 by the very brave and determined staff of the organisation that maintained it, the International Institute for Agricultural Research in Dry Areas (ICARDA), and placed in the Global Seed Vault on Svalbard. These collections, along with others around the world, are the bedrock of breeding research and development.

A Man Determined to Bake
an Ever-Better Loaf

This is where Andrew Whitley enters my story. It was a long, drizzly drive from Edinburgh, where he had picked me up, to his mill at St Monans, 80 kilometres (50 miles) north across the Firth of Forth, in the East Neuk of Fife. Already deeply involved in the Real Bread movement in the UK, Andrew founded Scotland the Bread in 2012 with his partner, the late Veronica Burke. They wanted to provide bread flour that was grown locally, would flourish in a low-input organic system, was healthy and could be grown sustainably. It was a simple mantra: 'Grow nutritious wheat and bake it properly close to home.' Andrew was determined to develop improved strains of wheat to produce flour to support a localised bread supply chain – from breeder to farmer to miller to baker to customer. He started to research varieties of heritage Scottish and Nordic wheats that were well adapted to grow in his corner of Scotland, ticking the boxes for resilience, being nutrient rich and delicious too. It took him five years before he was able to launch his first three fine wholemeal flours, developed, grown, milled and sold locally.

On the drive to his mill, I quizzed Andrew on his deep knowledge of local wheat varieties and his work in promoting bread made from Scottish grains. He was keen to point out that, even after more than a decade, the project is still only at the start of his ambition to transform bread production in the UK. He wants to change the entire system, with a fair deal for farmers growing locally adapted wheats, a shorter supply chain between field and baker, and a firm move away from ultra-processed factory-produced pap. It's a wonderful, inspiring ambition. The part of his story that interested me most was his work in the development and improvement of grains, both

Using One's Loaf

FVs and populations, that offer a sustainable and far healthier alternative to the high-yield, homogeneous cultivars that represent more than 85 per cent of the bread flour we consume.

It's a Small World

Andrew, a linguist who used to work for the BBC Russian Service before he became a baker, had discovered the importance of bread in Russian food culture when at Moscow University in the late 1960s. Travelling across what was then the Soviet Union, he found that the official currency exchange rate made price-controlled black rye bread pretty much the only affordable option for an impecunious British student. I made films there in the early 1990s, when the Soviet Union was collapsing, and, as someone who too had survived on this ubiquitous loaf when it was about the only thing one could buy, I can attest to its unique flavour!

Gene banks only make very small quantities of their seeds available for researchers and plant breeders to study and experiment with, so care must be taken not to waste these precious resources. Andrew, Anne Parry and others had been led to believe that the Hen Gymro accession known as S72 had traits that could be of more use than S70, which was being grown by a handful of enthusiasts. But the only place Andrew knew that held S72 was the VIR in St Petersburg; it had been added to their collection by T.J. Jenkin of the Welsh Plant Breeding Station back in the 1920s.[7] I had visited the VIR in the early 1990s. It was a troubled place then, with the priceless collections of more than 250,000 accessions of wheat and other cereals neglected and poorly maintained. Today it is in better shape and remains a key player in the conservation of global plant genetic resources.

The Accidental Seed Heroes

Might it be possible, Andrew wondered, to bring a sample of S72 back to the UK?

The number of growers, millers and bakers working on heterogeneous cereals for baking is small. They all know each other. One is Andy Forbes, a London baker, passionate about promoting traditional and FV wheats. He was growing small quantities in South London in 2008. He is an important part of the story of the restoration of traditional wheat into British bread and was keen to receive a few Hen Gymro S72 seeds from the VIR to test, trial and hopefully bulk up to increase the diversity of FVs he was experimenting with. Needless to say, his emails and letters went unanswered. Then, in 2014, Andrew Whitley made a visit to St Petersburg after having heard Professor Igor Loskutov, an expert on oats working at the VIR, give a lecture to a gathering of Nordic agriculturalists in Finland. Professor Loskutov was more than happy to give him seeds. Andrew received 10 grams (less than half an ounce) of the precious grains and handed the lot to Andy, who, with the involvement of an organisation he had helped to found, the Brockwell Bake Association, was able to bulk up seed stock, growing it first on allotments and community gardens in South London and subsequently with local farmers who could produce larger quantities for milling.[8]

At the same time as the Brockwell Bake Association was bulking up Hen Gymro S72, Andy Forbes suggested that the handful of members of another group dedicated to growing and milling traditional wheat varieties, the Welsh Grain Forum, of which Anne Parry was a founder, combine their grains with his to create a diverse Hen Gymro FV population. The grain that Anne mills in West Wales is grown by Mark Lea, who lives just over the border in Shropshire and is the main provider of seed stock to other growers in the

Using One's Loaf

forum. As I write, this unique population of Hen Gymro is being grown by an increasing number of farmers who welcome its ability to flourish in less than ideal conditions and provide reliable and consistent harvests of the most delicious and nutritious flour that I bake with every week! To be sure that the population mix can be properly maintained, all the different lines that make up the population are grown separately in small amounts by Ed Dickin at Harper Adams in Shropshire, one of the leading universities in the world specialising in agriculture and farming. As we shall see, maintaining diverse wheat populations and being able to refresh and adapt the mixes within these populations demands the disciplined involvement of plant scientists.

I, and all those shoppers who support bakeries making bread from locally grown and milled wheat, are beneficiaries of a wonderful and, I passionately believe, crucial form of public and private partnership. This is a marriage between agronomists and plant breeders at institutions like Harper Adams, the James Hutton Institute in Scotland (which has collaborated with Andrew Whitley), the John Innes Centre in Norwich and farmers, millers and bakers across the country. Between them they are maintaining, improving and developing diverse and resilient populations of wheat mixes, giving us healthy and truly delicious bread. The diverse mix of different lines of Hen Gymro is something uniquely Welsh that is also now being grown beyond the Welsh borders in England.

The YQ Story

The development of populations of bread wheat may have started gently in the UK at the dawn of the twenty-first century, but it is now a global revolution. The approach to modern plant breeding, which applies as much to wheat as to other crops, is focused around providing new cultivars that address specific requirements, driven first and foremost by entirely understandable and, for commercial reasons, essential qualities: yield, and resistance to pests and diseases. As with other crops we have discussed, the focus of the breeder is on identifying specific genetic traits and the genes associated with them to create something that is distinct, uniform and stable (DUS), and over which the breeder controls the intellectual property of the seeds. This means that every plant in a field of modern wheat is homogeneous – genetically the same, equally and highly productive with identical traits and morphology. The plants look the same and behave the same because they are the same! This type of crop is therefore highly vulnerable to changes, whether biotic or abiotic, that it has not been specifically bred to resist; extremes of weather and pathogens that have adapted and evolved to survive and thrive in these monocultures being just two.* Modern wheats do not do well in soils that have not had additional inputs of artificial fertilisers. Like Ferraris, they need plenty of fuel! Diverse wheat populations that have been bred or selected specifically to thrive in low-input agroecological farming I find an altogether more exciting proposition.

The opposite to a field planted with homogeneous cultivars is one that is heterogeneous; in other words, made up of genetically diverse varieties. They can be pure FV populations as we have seen with Hen Gymro, or a mixture of varieties that have been developed and improved by inspirational

---

* Biotic stresses are caused by other living organisms, whether harmful or beneficial; abiotic stresses come from the effects of natural events such as flood, drought or extremes of temperature.

Using One's Loaf

plant breeders like Anders Borgen from Denmark, whom we meet later in this chapter. And then there are traditional varieties that have been mixed together and allowed to evolve naturally – evolutionary populations – and this is where YQ comes into the story.

Martin Wolfe (1937–2019), who laid the genetic foundations for a new approach to creating heterogeneous wheat populations, was for twenty-eight years Professor of Plant Pathology at the Plant Breeding Institute (PBI) in Cambridge. In 1997, after he had retired, he and his wife, Ann, established Wakelyns Agroforestry, a farm in Suffolk that integrated forestry with agriculture. Wolfe wanted to reintroduce genetic diversity in cereals. In 2002, he sowed a mix of twenty distinct old and modern wheat varieties because he felt each had traits that would complement the others. The tall and the short together would reduce the risk of failure in flood or drought, and the mix of varieties would provide better resistance to disease, as any infection would struggle to move across the whole crop. His selection included varieties with very good root structures, which would be able to scavenge for nutrients, unlike short-rooted elite cultivars that have to be spoon-fed nitrogen fertiliser: in other words, his seeds would grow well in low- or no-input situations.

Wolfe crossed all twenty varieties with each other, creating 190 hybrids that were mixed up and sown together the following year. This resulted in a genetically diverse population. The next year he sowed without any selection and repeated the process over the following seasons. He gave Andrew Whitley some of the crop to assess its suitability as a bread wheat specifically for small-scale artisan bakers. Working with the Organic Research Centre (ORC) as their principal scientific advisor, he studied the wheat's performance from one year to the next in various locations and soils, both organic and 'conventional-intensive', round the UK.

The varieties that Wolfe grew had been part of a breeding programme at the ORC and he wanted to create a population of wheat that was not reliant on the inputs conventional breeders require: nitrogen fertiliser, plus a suite of fungicides, herbicides, pesticides and growth regulators. Known at first as the ORC Wakelyns Population, its official name YQ, approved by the Department of Environment Food and Rural Affairs (DEFRA), is an abbreviation for yield and quality. Kimberley Bell of Small Food Bakery in Nottingham championed YQ as her go-to bread flour and it was then taken up by several artisan bakers. Twenty years ago it was illegal to grow and market YQ because populations of heterogeneous wheat do not conform to EU law, which demands that all commercial seed be DUS. A concerted effort by the ORC and others, including Anders Borgen in Denmark as we shall see, persuaded European officials of the benefits of heterogeneous wheat populations for organic and sustainable cereal production. In 2014, under an EU statutory instrument, permission was given, initially for four years, then extended for a further three, for YQ to be grown, exchanged and sold without having to adhere to DUS rules.[9] In 2021, when there needed to be further negotiation if DUS rules were to be waived, the UK had left the EU and the world was reeling from Covid. But by now DEFRA was better informed about the benefits of heterogeneous crops. A small group of determined growers and activists under the auspices of the UK Grain Lab was able to extend the permission for any heterogeneous cereals until 2030. It is hoped that the evidence that wheat populations like YQ present huge benefits for human health, biodiversity and food security will persuade the authorities to continue to make exemptions to DUS. Today, YQ is grown by an increasing number of farmers supplying bakeries across the country.

However, YQ is the result of crossing from a limited number of commercial varieties that were developed for modern agriculture in the last century. Its lower gluten levels have proven a challenge for bakers, who must employ all their skills to get the best out of the flour. I use YQ grown by Fred Price on his farm in Somerset for pastry. As we shall see, he is another important player among the cohort of a new generation of young farmers growing and trialling heterogeneous bread wheats.

Breeding for Better Bread

I was keen to learn more about what is involved in selecting, breeding and maintaining large numbers of different varieties of wheat and lines of FVs. So, I needed little excuse to travel to Denmark to talk to one of the world's most respected breeders and advocates of traditional wheats, Anders Borgen. It was important for me to meet him and learn about his approach and philosophy as a plant pathologist turned breeder, because it was his introduction to YQ that started him off on a different approach to developing wheat populations.

North of Denmark's second city Århus, the barn was lit from a low autumn sun casting shadows across the orderly boxes filled with hundreds of bags and sacks of seeds – mostly wheat, but rye and spelt too – which took up much of the floor space. Many of the smaller bags contained just a handful of grains, each with a large label on a stick firmly planted in it, bearing details of the dozens of lines Anders had selected as part of his breeding and improving programme. He was pleased to show me some of the machines he uses to manage seeds. One is able to sort according to colour; another, shape. Anders uses infrared scanning to determine the amount of protein in each seed and has a tiny kneading machine that can, with just a teaspoon of dough, determine the levels of gluten in any of the hundreds of lines he grows every year.

Anders is best known for his work, started in 2007, on a breeding programme focusing on resistance to a nasty disease, stinking smut – so named because when the spores are released at harvest time they give off a fishy smell. Better known to me as common bunt, it is caused by two closely related fungi, *Tilletia tritici* (syn. *T. caries*) and *T. laevis* (syn. *T. foetida*). It is a disease that can have devastating effects on both yield and quality. Anders' work on developing resistance to common bunt, which is a serious management issue for organic growers, has helped enable a dramatic increase in organic production in Denmark, inspired by YQ.

Every year, Anders selects from the harvest healthy seeds that show diversity and, above all, good culinary qualities. Unlike Martin Wolfe and his YQ, which is a population wheat that evolves naturally, Anders carefully selects from individual strains and lines for traits he can include in highly diverse mixtures and which can be maintained from year to year. Anders is developing a range of cereals that can be grown and further developed across Europe and beyond. Along with growers and breeders passionate about our health and the importance of diverse cereals in our diet, they see a way forward for farmers. One of the varieties he has bred is Mariagertoba, a hard, high-gluten spring wheat, which is prized by many bakers. It resembles a generic red Canadian variety that originated in Manitoba and is marketed under that name. Anders' wheat is made up of a mixture of fifteen varieties, which, as well as being perfect for bread-making,

Using One's Loaf

express excellent resistance to a host of pathogens including common bunt, mildew and rust.

Fred Price is growing Mariagertoba to supply the bakery on his farm in Somerset. The Organic Research Centre is using a few acres of his land to trial both modern elite cultivars and populations, with a view to establishing a viable seed production network to supply the growing number of farmers keen to move into a more sustainable business model. Fred has no control over the weather, nor can he set a price for his wheat, which is determined by international traders and a sophisticated futures market that provides no security and narrow margins – if any – to the farmer. So, ever-larger farms that can benefit from their scale to reduce costs alongside a pursuit for ever-greater yield become the key driver in deciding what to grow, and it is the boast of giant seed companies that only their cultivars will deliver. As Fred found to his cost, to increase yields he needed to apply ever-greater quantities of fertiliser and chemicals. As he told me, the latest cultivars are the equivalent of an iPhone, supporting an industry that relies on built-in obsolescence to keep their farmers on the high-yield treadmill. So he side-stepped the commodity system and focused on growing what he wanted and what he believed the consumer wanted too. Yields were low to start with, but as he was getting a much better price for his crop by selling within a closed community made up of himself, a miller, a baker and customers, he was no worse off, and his heterogeneous wheat populations have adapted and improved in the last few years. An agroecological approach to farming has stopped the degradation of Fred's land. Now it is healthier and more productive and his overheads are greatly reduced. Taking the best of systematic plant breeding and generations of knowledge into traditional, resilient ways of producing crops, Fred seems to me very close in both spirit and intent to farmers I met in Ethiopia, who too are dependent on wheat and indigenous cereals to support their families and communities.

What Goes Around, Comes Around

In Addis Ababa I met up again with Dr Basazen from the Ethiopian Biodiversity Institute (EBI), who told me a remarkable story about the power of diversity and resilience in locally adapted elite crops. Early in the twenty-first century, the Ethiopian Institute of Agricultural Research (EIAR) had developed a new rust-resistant wheat cultivar, Enkoy, that could flourish in low- to middle-altitude regions with little rainfall. However, after a few years, a rust pathogen evolved through mutation to flourish on Enkoy. Farmers stopped growing it and switched to a less elite variety that expressed resistance to the mutated rust. Today, Enkoy is back in production because the race of stem rust it was susceptible to has disappeared due to a combination of climate change and Enkoy's absence as a host.

To help me understand what had happened, Dr Basazen explained the 'gene for gene' concept. What geneticists know is that for every gene that confers resistance to a pathogen there is a corresponding gene that confers pathogenicity (the ability of an infectious agent to cause disease in a host). When farmers switched away from Enkoy to growing a different wheat cultivar, the rust had no host and went extinct. Now Enkoy flourishes again, although it will probably only be a matter of time before a new race of rust mutates and becomes problematic. Then the whole gene-for-gene cycle can repeat itself. Hence the benefit of diverse populations, where the mixtures include varieties with a range of traits and disease resistance that acts as an insurance against total crop failure.

A Grain for All Seasons

One of the main reasons I wanted to visit Ethiopia and listen to its indigenous farmers was that the country is one of the world's most important centres for crop biodiversity. We depend on biological diversity for our very survival, which is why doing everything we can to halt its degradation and with it the extinction of priceless genetic resources is of existential importance for all of us. Sorghum (*Sorghum bicolor*) was first domesticated in Sudan about five thousand years ago, making it one of the earliest of all cultivated crops in the Horn of Africa.[10] There are about twenty-five species native to Africa. It's the fifth most cultivated cereal in the world and one of our most important cereal crops – and not only in this part of the world. I have enjoyed eating sorghum chapatis in India and fed it to my chickens at home from seed imported from the US, which grows more of this amazing cereal than any other country.

Like all of the native and indigenous crops of Ethiopia that have been maintained and improved by farmers, sorghum is drought tolerant, highly nutritious, can be harvested multiple times from a single sowing and stores well. In a number of regions in southern Ethiopia, I saw it being harvested and resprouting for a second and even a third crop. The plants can grow to 3 metres (10 feet), which makes them an important animal feed too; everywhere on my journey I saw great stoops of sorghum straw, as tall as the homes they were propped up against, in readiness to feed families' goats and cattle in the dry season. When not being eaten – as with Hen Gymro closer to home – the straw would be used for thatching. The markets were well stocked with sacks of colourful grain: white, red and yellow, evidence of the genius of Ethiopia's farmers in selecting and maintaining a great diversity of locally adapted FVs.

It's All in the Colour

Traditional farming has, for hundreds if not thousands of years, grown populations of crops. These exist in two forms: maslins, which I expand on later, where different species such as wheat and barley are grown together; and varietal populations of a single species. Just to confuse matters, some breeders of population bread wheats like Anders Borgen might include other species, such as rye or spelt.

The sorghum farmers of Ethiopia are growing varietal populations, but because the phenotype of each variety is easy to identify and sorghum seed heads, which are large, are harvested intact before threshing, it is easy to separate out each member of the population to create a new mixture for the following season. Research at the end of the last century confirmed that the selection criteria made by farmers was maintaining FV diversity even when high-yielding modern varieties (HYVs) were grown in the same environments.[11] Modern plant breeders have used traits found in FVs to develop HYVs – most evidently in the US, where sorghum is an important biofuel as it grows well on marginal land, requires little input and, unlike its maize equivalents, is drought toler-ant.[12] Like wheat and a number of other cereals, sorghum is self-fertile, so FVs are the result of accidental mutations and rare crossings that are retained by savvy farmers.

Using One's Loaf

Ethnobotanist Dr Alex McAlvay, from the New York Botanical Gardens, told me about his observations of farmer selection of sorghum in Ethiopia. Average populations are mixtures of up to twenty-five FVs. Birds can be a big problem, but some sorghum, identified by its deep red seed heads, like the ones I saw being carried through a village in the Konso region, is bitter, so it is unattractive to birds but good for brewing. Farmers grow it around the margins of their fields to put the pesky varmints off – and it works. Some of the varieties in the mix show resistance to locusts but not to fungal infection. Others with different-shaped heads show resistance to fungi but not to insects; still others, mostly white, are delicious and nutritious, but birds love them. By growing these mixtures together farmers ensure they have resilience in their cropping. The selection criteria they employ also include optimising the mixture for yield, total biomass (how much stem and leaf are produced for fodder), market value, milling and brewing quality, and how long it takes to ripen. The women will select the best grains for cooking and those the animals will eat. Only elements of the crops are susceptible to different biotic and abiotic pressures, which results in reduced disease and attack from pests and a reliable supply of delicious seed.

There are limits to how much molecular breeding can be used to imitate the physiological trade-offs indigenous farmers make in their selections. Because HYVs are genetically narrow, the result of only a few selection criteria for high yield in favourable conditions, it is impossible for geneticists to create a single type that tastes great, is unloved by birds, is unattractive to bugs and locusts and has resistance to fungal infection. Leave these HYVs to US growers in their giant monocultures by all means, but in Ethiopia and other food-insecure countries, the development model for sorghum needs to be a marriage between farmers and scientists employing evolutionary breeding approaches. A good example of this can be seen in the development of improved resistance to the effects of the parasitic plant witchweed (*Striga hermonthica*). Pretty as its purple and pink flowers are, the roots of this plant attach themselves to sorghum and can cause 100 per cent crop failure. Sorghum is inter-fertile, meaning wild and domesticated species can cross-pollinate. There are three main genotypes: cultivated, wild and cultivated-wild cross breeds. The domesticated genotypes are the most susceptible to striga. Recent research has shown that certain wild sorghum accessions, along with one FV with the exciting name N13, show excellent resistance to this nasty parasite.[13] With this knowledge, geneticists use marker-assisted selection to identify which parts of the genome of wild sorghum should be introduced into new cultivars through both classical and molecular techniques. Breeding striga-resistant cultivars has been championed by Professor Gebisa Ejeta, who won the World Food Prize in 2009 for developing the first elite hybrid sorghum that ticks a lot of boxes: it is not only resistant to striga, but also shows excellent drought tolerance. It is estimated that striga affects over 60 per cent of farmland under cultivation in sub-Saharan Africa and that 300 million farmers are suffering yield losses totalling more than $7 billion, so improvements in breeding can have a dramatic and positive impact.[14]

Farmers integrating a suite of elite HYVs into traditional populations and maintaining them would be expected to improve yield. According to Dr McAlvay, the evidence

Using One's Loaf

to date is mixed. Some farmers, particularly those who are involved in research programmes or whose land is adjacent to demonstration sites, are growing HYVs. However, being pragmatists with the knowledge of countless generations of subsistence farmers before them, they prefer to grow the elite varieties in separate plots along the field margins. One of the limitations of these HYVs is that they are much shorter than FVs, so do not fit well into an existing population. Nevertheless, in drought years farmers turn to them because they ripen sooner. Sometimes, however, HYVs are grown, harvested and stored together with the FVs, and threshed together or separately depending on how they are to be used.

There's More than One Type of Population

When I spoke with ethnobotanist Professor Zemede Asfaw at Addis Ababa University, he pointed out that a policy of monoculture that puts yield front and centre of food production is short-termism: people are starving, and providing food aid is just firefighting. Indigenous crops are needed to maintain resilience, and these include the cereal mixtures known as maslins, which I mentioned earlier. These have been cultivated in Ethiopia and Eritrea for millennia and continue to be grown in Georgia and on some Greek islands. In northern and eastern Ethiopia, growing wheat and barley together is an important subsistence strategy for many farmers, because they see greater resilience against pests, diseases, flood and drought than with either of the crops grown separately; when wheat and barley are sown together, the barley will survive drought, ensuring the farmer can harvest a crop for survival. In a normal season the more valuable wheat yields more.* To some extent maslins imitate the diversity of species found in wild grassland and savannah, and they are especially adapted to low-input farming systems. Yet they are in decline in Ethiopia because of a variety of factors, including government pressure to grow elite wheat monocultures, more effective irrigation and the use of artificial fertiliser. Ethiopians would appear, from what I saw them eating, to have fallen in love with highly processed white bread made from imported modern wheat cultivars. These are replacing diverse agricultural systems that had been growing maslins. Winning back traditional food culture requires better education and understanding of its deliciousness and nutritional superiority.

Researchers have found, from talking with farmers, that maslins' yield is superior to that of individual species when grown on their own; they show greater drought tolerance, as well as pest and weed resistance.[15] All types of cereal mixtures frequently show greater yield stability, with better quality and more nutritious flour for bread-making.[16] Maslins grown across southern Europe and the Indian subcontinent also include sorghum and rice; field beans and peas; wheat and rye; barley and oats – all mixtures where one species will survive in a climate crisis when the other fails.[17] It's an insurance policy for food security that should be emulated and used to replace the intensive and environmentally unsustainable cropping systems of which the Global North has made a specialism.

---

* Barley, unlike wheat with its quirky genome, can be grown worldwide, so offers viable solutions to diverse cereal production.

CHAPTER FOUR

Setting Seeds Free

Plant Breeding for an Equitable Planet

*Plant breeding is the art and science of chang-
ing traits of plants in order to produce desired
characteristics and it can be accomplished
through many different techniques ranging
from simply selecting plants with desirable
characteristics for propagation to methods that
make use of genetics and chromosomes, to more
complex molecular techniques.*

Melese Lema Tessema, Researcher
in Ethiopian Agriculture

There is a wonderful apparent randomness about agroforestry – the practice of integrating trees into farming
systems for a variety of reasons – which manifests itself very
clearly when all the cultivation is terraced. In the Konso
region of southern Ethiopia, at least fifty types of food crop
are grown together: vegetables, cereals, fruits and nuts. Many
of the walls of the terraces are strengthened, with introduced
species including mango, guava, avocado and indigenous
coffee trees growing on them. It is not just food crops that

The Accidental Seed Heroes

bedeck the terraces, but trees with other important uses like the Abyssinian or African milk tree (*Euphorbia trigona*), which is used as a fire extinguisher in the event of stubble burning getting out of control. A common native tree is the locally named *oybata* (*Terminalia brownii*), which is used in construction; both bark and leaves are important as animal feed and medicine – they're also good for healing cuts. But it is within the terraces themselves that the effectiveness of agroforestry becomes most apparent through a kaleidoscope of crops in different stages of development. From emerging seedlings to clumps of sorghum awaiting harvest; drifts of shoulder-high khat ready to have their leaves plucked; mango trees laden with ripe fruit waiting to be grasped and savoured; scrambling sweet potato plants rippling along the ground like the botanical equivalent of an express train. The sheer diversity took my breath away, along with the understanding from the people I met that no one in the region was going hungry.

I was in the village of Mecheke, deep in the heart of the world heritage site, the Konso Cultural Landscape of Ethiopia's southern highlands, at the start of the short rainy season in March. It is usually the men who do the planting, but as I walked among the terraces and saw the amazing diversity of crops being planted at that time, it was a young woman who showed me what she had sown a few days before: maize, and black and red beans. She had randomly spread the seeds among already growing different-coloured FVs of sorghum, yellow, green and white pigeon peas (*Cajanus cajan*), barley, wheat, cabbage and khat (*Catha edulis*). Khat is a deep-rooted plant that helps bind soil, so is very useful in a diverse cropping system. In some parts of Ethiopia, it is replacing coffee as a cash crop because it grows quickly and can be continuously harvested. Its sole use is as a narcotic; chewing it – something

76

Setting Seeds Free

that has been practised across eastern Africa for thousands of years – induces mild euphoria, sociability and loss of appetite.

Throughout my travels in southern Ethiopia, meeting growers, botanists and agronomists, I was constantly made aware of the pragmatism and resourcefulness of the people. Before I arrived in the country, I had been led to understand from some environmental activists and ethnobotanists that Ethiopia was compelling farmers to abandon their local varieties and grow imported modern hybrids of wheat, pulses and maize. Nearly all the farmers I met were indeed growing crops using improved and modern cultivars they had bought either from government seed distribution centres or from local dealers. Many of the younger ones were growing only cash crops such as modern tomato and pepper cultivars. This was for good reason. They delivered greater yields than traditional FVs and fetched better prices in the market. But this didn't stop older farmers from also growing and maintaining seed from their own, locally adapted varieties. In fact, they and their families preferred to eat crops from their own saved seed because they tasted better. They would sell any surplus in the markets alongside the modern stuff. Their love of these varieties was clear to see: I met four generations of farmers in one field and all of them had every intention of continuing to grow, save the seed and maintain the diversity of their crops. These farmers' knowledge was hugely impressive and hardly surprising; as soon as children could walk, they would help on the land – often herding cattle and goats and driving donkey carts, but also gleaning, weeding, planting and learning.

Embracing a Holistic Vision

Subsistence farming is not for the faint-hearted. When everything you rely on for survival is dependent on the rains arriving on time and with absolutely no margin for error in agricultural practice, literally living off the land tests the people every single day. Yet, unlike many British farmers I know whose default is generally to grumble because something is always going wrong – be it market prices, loss of subsidy, the weather, the banks, the government or the customer – I have found indigenous farmers at peace with themselves and the realities of their lives, despite their many hardships.

The people of Konso kept their thatched grain stores topped up with sorghum year after year. This is one reason why they have avoided famine over so many centuries. With such diversity of crops under cultivation, regardless of the vagaries of the climate, there was always something to eat. Pigeon pea, indigenous to India, would be left to grow over two years, the tall, woody plants expressing excellent drought resistance. Many of the crops were grown as populations – different varieties growing together: fava beans, common beans, maize, sorghum, wheat, barley, even garden peas. I saw locally adapted perennial kale growing both on the edges and within terraces, cut back in the long dry season and allowed to resprout with the rains. The locals considered it a useful crop, but not as delicious as proper Ethiopian cabbage (*Brassica carinata*). I can relate to this sentiment because there is only one kale I truly love – asparagus kale – but I suffer lesser varieties when they are all that is on offer in my garden in early spring.

Konso is an inspiring example of a way to build resilience into a food system that has been proven to last for centuries. The densely cultivated landscape I explored, rich in crop diversity; a mix of indigenous and introduced varieties; FVs and elite crops that are continually evolving and growing together – they all go to demonstrate the enduring power of diversity in adaptive food systems that will feed us over the coming decades.

Setting Seeds Free

From Feast to Famine

Before the birth of agriculture some twelve thousand years ago, humankind were hunter-gatherers, depending for their food on what was readily available. Plants provided all the seasonal fruits, roots, leaves and seeds for a healthy and sustainable diet. Although our ancestors could identify hundreds, indeed thousands, of plants for food and medicine, they probably only relied on a relatively small number of 'staples' for their daily needs.[1] There is no evidence that these pre-agrarian communities ever suffered from famine, although population growth and pressures on food supply were almost certainly two of the key drivers of settled farming and domestication. If a regular and favoured source of nutrition was not available for whatever reason – drought, flood or changes in climate – there was always something else to be had. The very few remaining hunter-gatherer communities in the world today are testament to a resilience that trumps the fragility of agriculture.

As soon as we started the process of domestication – that is, selecting only for traits that suited settled agriculture – humans became vulnerable to starvation for the first time. The history of famine is a brutal one. The reasons for it are many, but until relatively recently climate change was not one of the main causes. Circa 2200 BCE a dramatic but short-lived change in the weather spelled disaster for the Egyptians. Some argue that the resulting famine could have been lessened if the rulers at the time had taken steps to make their food system less precarious – a single harvest failure could lead to famine and social upheaval – by holding greater reserves. Nonetheless, disease (often exacerbated by an associated famine) had a far greater impact on society than natural disasters and was usually caused by human miscalculation.[2] But with civilisation came war and food

The Accidental Seed Heroes

– or the lack of it – as a weapon. It is conflict more than anything else that precipitates famine. Also, the manner in which food was shared, with elites benefiting at the expense of the poor, has resulted in the deaths of countless innocents since the dawn of civilisation. Until the advent of modern plant breeding, famine and disruptions to food supply were never due to a lack of diversity in crops. And then there is climate change. Our best insurance against the vagaries of an unreliable climate is crop diversity, moving away from the genetic narrowness of the limited number of crops we grow.

To give some context here: of the three hundred thousand species of edible plants in the world, only about two hundred represent the total diversity of what we eat, and of those just thirty supply 95 per cent of all our food. Three plants, all grasses, provide more than 50 per cent of all the calories consumed in the world: they are maize, wheat and rice.[3] In the last one hundred years plant breeders have been consistently narrowing the genetic diversity of these crops so that they can function within a highly mechanised and industrialised food system. If you thought that only ancient civilisations screwed up on building secure agricultural systems, think again. What is happening now with modern plant breeding – despite all the marketing hype – is not a fix as our climate continues to warm and our soils continue to degrade. Learning from and avoiding the mistakes of the past is no longer a choice: it's a necessity.[4] The seeds we sow and the ways we cultivate them are fundamental to our food security.

Monocrop Madness

The Dust Bowl of the 1930s – where agricultural land across the entire US Great Plains region turned to dust as the result of a catastrophic failure in farm practice and land management, exacerbated by drought – is an event that has seared its way into American literature if not memory. *The Grapes of Wrath* by John Steinbeck – a book that I first read in my teens – demonstrated in the most brutal way how the sheer stupidity, ignorance and hubris of a generation of farmers and their political masters were in part responsible for the ecological and agricultural destruction of the American and Canadian prairies (a drought lasting up to eight years being the other major contributing factor). It needn't have happened, even with the drought. Practising farming methods that were adaptable to periods of low rainfall, known as dryland farming, would have prevented the soil erosion and degradation caused by the wind that destroyed the topsoil in the region. A sensible return to grazing and sowing a greater diversity of crops, rather than concentrating on wheat, would have had a restorative effect on the health of the land. But this did not happen in any meaningful way. Instead, high-input techniques won out: digging deep wells to extract water from the aquifers beneath the plains, using more pesticides, herbicides and fertiliser as well as greatly increased mechanisation on huge farms certainly had a positive impact on yields, regardless of the amount of rainfall. Technology had provided the solution and the bad days of the Dust Bowl were consigned to history... Or so they thought.

And Then It Was Maize

As the Dust Bowl crisis clearly showed, agricultural practice that ignores the need to nurture and care for the soil and to build diversity into farming methodology ends in disaster. In the 1930s the crop in question was wheat; today it is more likely to be maize that is grown almost exclusively using homogeneous, genetically modified cultivars. A great critic of the present system of global food policy is the American sociologist Jack Kloppenburg. In 2005, he presented clear evidence of the dangers of monocropping of homogeneous maize when a highly infectious fungus, Southern corn leaf blight (*Bipolaris maydis*), resulted in a catastrophic drop in yield amounting to – in today's value – about \$7 billion of losses for farmers. Kloppenburg and others linked this disaster to hybrid cultivars, but there was academic blow-back from some who took great exception to his and others' views.[5]

As renowned environmental scientist Jonathan Foley so eloquently wrote in *Scientific American* over a decade ago, maize cultivation is seriously vulnerable to shocks, whether triggered by the weather, pathogens or economic factors.[6] Because maize is grown for more than just human consumption – it is a bedrock of global animal feed and a key ingredient in bio-fuels – any loss of production reverberates across world economies.

According to studies by NASA, climate change is likely to cause drastic reductions in global maize yields, with projected declines of 24 per cent by the end of the 2020s because some tropical regions will be too hot for maize cultivation. Interestingly, global warming isn't necessarily bad for all crops; NASA predicts wheat yields could increase by 17 per cent in the same timeframe because higher levels of CO_2 in the atmosphere have a positive effect on that plant's growth, but with heat and drought it may prove otherwise.[7]

Standing Up to the Powerful

It seems that there is an unbridgeable gulf between the opposing views of the world's largest biochemical company Bayer and its critics. Bayer says its products are crucial tools to deal

Setting Seeds Free

with climate change and to de-carbonise the world. To influence food production in the Global South they are already beginning to deliver seed and technology to one hundred million smallholders; this programme is due for completion in 2030 in Africa, South America and Asia. I believe Bayer is systematically planning to take away choice and control from said one hundred million, who will be dependent for their future success on using patented seeds and inputs from one source. What will happen to the countless amazing, diverse, resilient and culturally important FVs and locally adapted varieties of which those one hundred million farmers have been custodians for generations?

When I challenged Bayer's media relations spokesperson Alexander Hennig on this, his replies begged more questions. He assured me the company would not *intend* to enforce its intellectual property (IP) rights against smallholder farmers who used farm-saved seed privately and non-commercially to escape extreme poverty. But it made me wonder just how he thought a farmer and her or his family could survive and flourish without growing crops they could sell! Hennig admits that farmers can save hybrid seeds but will lose the increased yield and other specific traits that have come about as a result of the breeding improvements known as hybrid vigour.

Many seed companies like to propagate a narrative about plant breeding that claims hybrids and genetically modified cultivars are superior to FVs, open-pollinated and traditionally bred varieties. But there is plenty of evidence to suggest that this is not the case. In particular, crops that are bred or selected to flourish in low-input environments invariably outperform homogeneous cultivars, which perform very badly without vast amounts of input. Despite acknowledging the importance of wild relatives and indigenous collections

83

The Accidental Seed Heroes

in supporting improvements in plant breeding, the business model now is for seed companies to own the IP and thus all elements of a plant's genome, paying lip service to the indigenous farmers who have been custodians of these genetic resources. What was once a common good to be shared for the benefit of all now becomes a company's property, with profits directed towards the owners and shareholders.

When responding to my questions regarding 'empowering one hundred million small farmers', Hennig pointed me to Bayer's 'Better Farming Life' venture, which has the headline: 'Smallholder farmers are the heartbeat of global sustainability... the world's most powerful agents of transformation. Only by placing them at the forefront can we hope to build resilient food systems and ensure the long-lasting health of our planet.' A noble sentiment against which no one can argue, but one of the first case studies Bayer trumpets is its training programme in India for safer use of chemicals in crop protection. A photograph shows farmers dressed head to toe in protective clothing, preparing to spray their crops with poisons.* I have witnessed countless examples of farmers employing agroecological and regenerative approaches to growing crops that don't require them to dress up like astronauts and cover their fields in expensive toxic inputs.

GM cultivars with genes inserted that are designed to make the crops immune to bugs or herbicides have vulnerabilities. Over a relatively short time – often as little as five years – the targeted pathogens or plants will themselves mutate and

---

* This image can be found here: https://www.betterlifefarming .com/our-stories/experimental-study-in-india-confirms-increased -awareness-and-high-adoption-of-responsible-use-practices-by -smallholder-farmers-after-bayers-safe-use-training.

evolve immunities, ensuring that the breeder must forever be developing new strains with different genes inserted to, hopefully, prevent said bug or poison from killing the crop. Ultimately, the bug wins and more genetic modification is required to continue on the breeding treadmill. We have no idea how the GM genie is going to impact us once it is in the public domain, but rather than agonise over what could happen I prefer to focus on what we can do without the need for Bayer and its ilk.

Evolutionary Plant Breeding – A Way to Restore Genetic Diversity

As a counter to the domination of the major seed producers, examples of farmer-led seed production and improvement are happening in many countries.[8] Farmers are also taking the law into their own hands, with some accessing pirated and illegal imports of GM seeds. This is particularly the case in India, as I explore in chapter 10, where it is happening with a number of crops, including GM rice and cotton.[9]

Research literature asking, 'How do we change the model?' is legion, pointing to the opportunities and benefits of evolutionary plant breeding (EPB) to build a sustainable and secure future for global food production. EPB is an approach that enables field crops to evolve through natural selection pressure and adapt to their changing environment. Traditional farmers have employed this method since the earliest days of domestication, but today there are more and more examples from around the world of evolutionary-participatory plant breeding (EPPB), where farmers, working with plant scientists, have selected varieties of many cereals, including rice, wheat and barley, from their own home-saved

The Accidental Seed Heroes

crops for ongoing breeding. The effect has been for farmers to create 'smart crops',* so called because they have been shown to have positive effects on human health, the environment and farm incomes.[10]

Evolutionary populations (EP) offer even more. A considerable body of research over the last sixty years indicates that diverse heterogeneous mixtures show greater resilience against pests and diseases. This has had a number of very beneficial effects for farmers: less dependence on chemical inputs and more reliable and stable yields.[11] As EPs evolve, their ability to resist many diseases increases as a result of natural selection. Vulnerable individuals in the population become fewer and those best able to resist disease increase. As long ago as 1961, research showed that consistency in yield is better in EPs, and is also greater than in conventionally bred mixtures, which themselves are better than homogeneous cultivars.[12]

Another great advantage of EPs is the speed with which they evolve and respond to changing stresses when growing in different places. This is known as an adaptation strategy. Just how quickly can they adapt to a new environment? From a purely scientific view, this speed is dependent on how much genetic diversity, which is associated with adaptation and the heritability of those adaptive traits, already exists within the population. Recent research with wheat has shown evidence of divergence in populations in different locations happening in just five years, with observable changes in the number of days before a plant reaches maturity occurring within ten years.[13] However, a

---

* The term 'smart crop' has been appropriated by the major plant breeders as a way to describe homogeneous cultivars – dumb crops, in my opinion – the polar opposite of evolutionary plant breeding.

Setting Seeds Free

crop's ability to adapt is still little understood and there is an ongoing debate about just how adaptive even the most heterogeneous crops are. In my own experience of selecting to make my veggies ever happier growing in my corner of southeast Wales, I have found that many varieties I have been saving seed of from year to year are more robust, earlier and better able to cope with the vagaries of the local climate. I am no scientist, but it seems self-evident to me. However, more research in this field is badly needed.

As we now know, one of the ways climate change affects human health is how it impacts agriculture. Two strategies to cope with our changing climate are firstly mitigation, which relies on the reduction of greenhouse gas emissions; and secondly adaptation, growing crops that adapt to climate change. EPs display precisely these two strategies: they require far fewer inputs than homogeneous modern cultivars and are wonderfully adaptive. And as the authors Salvatore Ceccarelli and Stefania Grando wrote in a paper published in 2020:

> As they [EPs] evolve, they generate a continuous flow of novel cultivated agro biodiversity both in space (because of divergent evolution in different locations) and in time (because of evolution within the same location) even within the same crop, which will be beneficial in increasing diet diversity and ultimately human health.[14]

The Long Arm of the Law

Frustratingly, when it comes to making EPs more widely available, there is a fly in the ointment: seed laws. A perfectly understandable need for growers to know that the seed they buy is what it says it is on the packet has resulted in an unholy alliance between governments and the major seed producers. Seeds that are produced through EPB do not meet the stringent requirements demanded of DUS, which could ultimately threaten their business model. But change is afoot. European legislation is now enabling some farmers to embrace EPB without breaking the law. In 2014, the EU recognised that some of the projects it was financing involved heterogeneous material that had no legal status. The European Parliament took a surprisingly pragmatic view and allowed:

> A temporary experiment at Union level for the purpose of assessing whether the production, with a view to marketing, and marketing, under certain conditions, of seeds from EPs of oats, wheat, barley and maize, may constitute an improved alternative to the exclusion of the marketing of seeds not complying with the requirements.

After a decade of trials involving breeders and farmers a conversation is beginning about what and how we should be breeding and growing our arable crops. As I explored in chapter 3, the British government has followed suit and allowed growers of heterogeneous cereal populations the same freedom – at least until 2030.

As a result, more farmers are able to grow and sell their seeds, with consumers, bakers and food businesses enjoying and benefiting from a mixture of deregulation and reinterpretation of the law. This is of particular benefit to the organic sector, which seeks to grow crops that work best in low-input systems. The development of cultivars expressly for use in regenerative and agroecological farming is at the heart of this 'experiment'. The process of selecting germplasm for organic seed production is focused on which breeding lines naturally survive pressures from pests, disease and changes in climate. Resilience and robustness are found in those seedlings that flourish without additional inputs of fertiliser and pesticides. One of the many things I have found so energising is the sheer number of organisations and institutions that are working globally in participatory breeding programmes to address the challenges across a number of crops.

An Indigenous Super Cereal

I have been fortunate to travel widely across Africa in my lifetime and I cannot remember a more interesting and diverse diet than in the country that has already featured prominently in this book: Ethiopia. Deliciousness in every meal and mostly eaten with injera, a dinner-plate sized fermented flatbread based on the great indigenous food, tef (*Eragrostis tef*). This tiny but hugely nutritious grain – a member of the same family of grasses as wheat – was one of the earliest crops to be domesticated in Africa some six thousand years ago.[15] Today, it remains a staple despite a high price (it is very labour intensive at harvest time), which means that most of the injera I used to scoop up my food with was made with the addition of maize flour or sorghum, or occasionally millet.

Despite its price, one cannot overestimate the importance of tef as a cereal. It can be grown in over a quarter of all land under cultivation in Ethiopia, from sea level to 3,000 metres (10,000 feet) elevation, and accounts for nearly a fifth of all cereal production. It is fast-growing and fits well into many different cropping systems; it is a brilliant catch-crop – grown between successive sowings of a main crop – especially when long-season crops like sorghum and maize have failed due to bad weather or pests. Although it can be attacked by some pests and diseases, for the most part it is little affected.[16] Most Ethiopians consume it on a daily basis. It is the most important crop both culturally and economically for the millions of farmers who grow it. Tef comes in two colours, red and white. Interestingly, white is more expensive but less nutritious than the red variety. Both are gluten free, high in fibre and an important source of protein. It has often been described as a superfood and not only sustains countless millions of people but also stores amazingly well: in a dry place out of the light it will keep for years. Its straw is much valued as a fodder crop. And it is used to make the favoured spirit of this part of Africa, *arak'e*. A true all-rounder!

Tef straw is used in traditional mud-and-daub constructed buildings. Seeing houses made using timber poles and cob, just as my family home in Devon was with clay, manure and barley straw, reminded me how little divides my relationship with the land from that of the people of Ethiopia, 80 per cent of whom (that's over 100 million) are involved in agriculture.

A Breeding Opportunity to Benefit Us All

As I travelled through the southern parts of Ethiopia's Rift Valley, I saw tef growing among other crops including beans, coffee and khat. In diverse cropping systems, especially with legumes, tef yields well without additional artificial fertiliser.

The challenge for plant breeders and farmers alike is that tef is prone to lodging – collapsing before harvest due to adverse weather – because it grows tall and has a shallow root

Setting Seeds Free

system. Windy and wet weather can cause havoc as harvest time approaches. Since the middle of the twentieth century much effort has been put into improving yield and reducing lodging using classical breeding, but with little effect. Until now tef has been considered an orphan crop and there has been minimal interest in applying molecular breeding approaches. Recently this has changed.[17]

Some tef is grown in South Africa and the US, but because it is indigenous to Ethiopia and only grown in small amounts in neighbouring Eritrea, there are limited quantities of germplasm grown outside these two countries that can enhance breeding programmes. Until now innovation has been through pure line selections of desirable traits found in the native FVs. Modern approaches using mutagenics and hybridisation with close relatives to create new cultivars less prone to lodging and with much higher yields are very challenging. There has been one successful white cultivar, Quncho, bred by Ethiopia's Debre Zeit Agricultural Research Center, which was released to farmers in 2006.[18]

Marker-assisted selection is proving a helpful tool since it started to be employed this century to better understand and sequence the tef genome. However, it is very difficult to identify the great diversity of varieties because their morphology is so affected by the environment in which they grow; plants of the same variety won't look the same when grown in different locations. Compared with other cereals, tef productivity is very low, 1.6 tonnes per hectare on average – wheat is three times more productive – but in the last few years progress has been made, with yields of some new cultivars more than double those of traditionally bred ones.[19] It may seem perverse to some, like me, who rail against the hegemony of the world's giant plant breeders, but the

The Accidental Seed Heroes

Syngenta Foundation for Sustainable Agriculture hosted by the University of Bern in Switzerland has established the Tef Improvement Project, which involves participatory breeding programmes with farmers.[20] The focus is on creating mutations and in vitro breeding to get round barriers that can arise from conventional crossing with wild relatives. Gene editing is also being employed as a research and development tool. How this might impact farmers' abilities in the future to save their own seed has yet to be determined. But today new lines with the desirable traits of reduced lodging, improved drought resistance and greater yield are sent to the Ethiopian Institute of Agricultural Research (EIAR) for further breeding, testing and passing on to farmers. Scaling up of these new cultivars has been taking place across the country since 2016, enabling subsistence farmers to grow varieties with more assurance of success and improved income.[21] This is an exciting example of indigenous farmers and plant scientists collaborating to enable greater food security, resilience and adaptability as the local climate changes.

Tef is also attracting considerable interest as a crop that can adapt to climate change and potentially become a major source of nutrition across the world. Some see it as an easy solution to a global food crisis, but I believe that, given time and further investment in research and breeding, it really can have a profound impact on achieving a carbon-neutral and ecologically sustainable future for food production. Above all, its place in Ethiopian food culture is assured and long may it remain so.

CHAPTER FIVE

A Future Full of Beans
A Solution to Save the World?

*Select the finest pods for seeds, let them remain
on the plant till the foliage withers, then pull
up the plants, and hang them in a dry place till
the pods are thoroughly dry.*

Thomas William Sanders,
Vegetables and Their Cultivation (1910)

The first bean I was introduced to when I arrived in Albania was a familiar-looking, large white specimen whose common name is the lima bean. True Albanian bean-lovers know it as *Trenare*. It is cultivated throughout the country and takes its name from the village of Treni on the Korçë plain in the southeast of the country. Trenare is considered by chefs and home cooks alike to be the best bean for the traditional dish of Tave me Pllaqi, or, as it is written on the menu of every restaurant in which I ate, Lima Bean in Tomato Sauce. My first encounter with that dish remains the most memorable.

Oda is a popular restaurant among locals in the Albanian capital Tirana. It takes up most of the first two floors of the owner's house. Each room is filled with tables, large and small, and I found myself sitting at one with a young couple who, when not gazing in rapture at each other, were intent on enjoying aubergines wrapped in vine leaves. As soon as I saw lima beans on the menu I knew what I had to eat and I was curious. The white lima bean is of the species *Phaseolus lunatus* subspecies *sieva* – the sieva bean – and I suspected that what I was ordering was a different and altogether more familiar species, *Phaseolus coccineus* – the runner bean. The dish I was ordering was probably misnamed when the lima bean was first grown in Europe five hundred years ago because, at first glance, one large white bean looks much like another and both species arrived here at about the same time.

Something else on the menu caught my eye – an Albanian classic: stuffed sheep stomach. Was this a Balkan equivalent to that stalwart of Scottish food culture, the haggis, I wondered? A glass of extremely drinkable red wine from the owner's vineyard on his farm just outside the city helped fuel my sense of anticipation. Oh, happy me! The stomach was stuffed with onions, herbs and tomato – about as far removed from a haggis as I could have imagined: meltingly tender, cut into bite-sized slices with a rich and piquant tomato and sweet pepper sauce. Slow-cooked bliss. Accompanying it came the 'lima' beans: glorious, meaty, soft and succulent. I was keen to know their provenance. My waiter said they had been 'imported' from Korçë – all of a hundred miles (160km) away. I could feel myself becoming besotted – not only with the food, but with the country. Sadly, he was unable to explain why it was called a lima bean but assured me that it was the same everywhere and to ask in other restaurants that had them on the menu where they had sourced them because imitations were to be avoided. He was also very excited to learn that the following morning I was planning to travel to Korçë, where only local beans would be found in

A Future Full of Beans

the town's restaurants. Leaving me to my meal for just a few moments, he then returned with a jam jar full of the dried beans. Would I like to grow these in my garden in the UK? You can guess the answer.

There were other FVs from southern Albania on the menu that evening, among them a climbing French bean with another wonderful name, Barbunjë e Blushit me Purtekë. Blush is the village associated with this variety, which translates as 'fresh bean from Blush with perch' – perch referring to the canes up which the plants grow. Fresh beans are also harvested for drying: another example of how traditional crops serve a multiplicity of purposes in the kitchen.

At Home Wherever They Grow

Just over the border in neighbouring Greece a prized local variety of large, white-seeded runner bean called *gigantes* is intrinsic to the cuisine. Across Europe I have collected and enjoyed eating local versions of the same species with national names including *haricot de soissons* in France and *judion* in Spain. These varieties have great cultural significance and are locally adapted FVs of *Phaseolus coccineus* that have been selected and maintained as a drying bean.

The shared sense of identity in the names of local varieties that I witnessed in Albania is something that I have also come across throughout Spain with the common bean, *P. vulgaris*. The Garrotxa region of Catalonia takes huge pride in honouring the diversity of its many FVs; nowhere more so than in the foothills of the Pyrenees where, it would seem, every village has its own, which it is happy to boast about. Claiming superiority over a neighbouring village's poor imitations is a matter of honour! On a memorable trip indulging in the region's superlative gastronomy I found myself wandering the narrow, winding medieval streets of Santa Pau. The fields all around were recently planted with beans, and in the shops bags of small, round Fesols de Santa Pau were for sale to local and tourist alike. (The French call this type of classic white drying bean *haricot blanc*.) Every restaurant in town had this bean on its menu, taking centre stage as a local delicacy and presented in a variety of guises. It's delicious, of course, with a distinct buttery chestnut flavour. It is acknowledged with Protected Designation of Origin (PDO) status as unique to Catalonia. And it is not alone. Just 6 miles (10 kilometres) west of Santa Pau is the larger town of Olot. It, too, has its own bean – *mongetes La Vall D'en Bas* – which takes equal pride of place on restaurant menus. In the same valley 12 miles (20 kilometres) further south, in the village of Rupit, the wonderfully named Mongeta del Ganxet also has PDO status. I bought seeds from a farmer who proudly grew them and wanted to be sure I knew exactly where they had come from by putting his name to the bags he was selling: *D'en Pere de Slica D'amunt* – from Peter above Slica. Slightly larger than Fesols de Santa Pau and oval in shape, these FVs have over the centuries become locally adapted to the unique and favourable climate and soils of the region.[1] Interestingly for me, all three varieties do pretty well in my Welsh garden and Fesols de Santa Pau is a staple in my larder, its organoleptic qualities undiminished by the change of location.

As with so many local and culturally important crops, maintaining production can be a challenge when fewer farmers are growing them. This is why these beans must continue to be celebrated by chefs and home cooks alike in order for them not to be lost. If farmers continue to grow them and diligently conserve seeds from the best plants for next year's crop, the beans will adapt to changes in climate through the wonder of evolutionary plant breeding. Having grown these beans for over fifteen years, I am in no doubt that they are well adapted to the soil and weather in my garden; they give consistently reliable harvests come rain or shine. They are diverse in their habit, climbing to a greater and lesser extent. I started to select and separate beans from the dwarfing and climbing plants, believing I could end up with two distinct traits. But every year, when I resow the dwarfing ones, some plants will start climbing and vice versa! So now I don't bother and treat the crop as a heterogeneous population.

A Possible Replacement

My quest for deliciousness is never dimmed. A couple of years ago I was asked by the UK's Heritage Seed Library to grow an American heirloom, the greasy bean. A classic of Southern US cuisine, it gets its name from the shiny appearance of the short, swelling green pods. There are as many ways to prepare this bean as there are cooks in the southern states; but most popular is probably the proper 'southern-style': sauteed in a skillet with garlic, bacon, butter and chilli flakes. Like all traditional beans, this one is dual purpose. When left on the vine to dry, the lumpy pods are known as 'leather britches' – which describes them rather well, I think. With summers now warmer, longer and later, greasy seems very happy in my garden. A huge cropper, it is now daring to replace Fesols de Santa Pau in my larder because it requires less soaking: just an hour or two from dry. After half an hour of gentle simmering it becomes a hugely versatile addition to many a dish.

Like its Catalan counterparts, greasy is small, white and round – another FV. The beans that populate the villages of Spain's southern Pyrenees all came originally from the same centre of diversity in Meso-America, brought to Europe some five hundred years ago. What excites me about all the FVs I maintain is their amazing adaptability and resilience. They not only provide us with protein, energy and fibre in our diet, but they fix nitrogen in the soil and require little or no additional input. As with all the varieties I grow, I keep back the earliest and best-looking beans to sow the following season – exploiting any changes in phenotype that are the result of natural evolutionary changes.

The Making of a Very British Bean

The small, round, white haricot bean that is used for canning is commonly known as the navy bean because it was first used to feed sailors in the seventeenth century. On the face of it, the ubiquitous tin of baked beans we see as we journey along the aisles of every supermarket in the country may seem to be an icon of British popular ready meals, but it was an American invention, the work of Henry John Heinz (1844–1919) in the early 1880s and his most successful tinned food. In 1886, he exported it to the UK, where it was first sold in London's smartest food shop, Fortnum and Mason. We Brits took to this little white bean in its sweet and salty tomato sauce, and early in the twentieth century everyone, rich and poor, was buying the now very affordable Heinz baked beans.

It was not long before Mr Heinz's beans were being processed in the UK. Today, 50,000 tonnes are imported every year, mostly from the US, to just one factory in Wigan, the largest of its kind in the world. It produces 3 million tins

A Future Full of Beans

per day, 2.5 million of which are consumed by my fellow citizens. According to a recent poll, 43 per cent of the UK population eat beans on toast at least once a week.[2] However, Heinz is not the only purveyor of beans in the country; Princes Group, a company established in Liverpool in 1880, is today part of the Mitsubishi Group and one of the largest suppliers of groceries in Europe. It produces 264 million tins of baked beans a year for the British market, importing between 100,000 and 110,000 tonnes of navy beans, mostly from the US. In 2023, for the first time, Princes undertook tests to make their baked beans from a new British cultivar, Capulet. It is the result of an exciting and potentially game-changing 12-year-long breeding programme that has been the brainchild of Professor Eric Holub, a geneticist at Warwick University's School of Life Science.

In a world where food security is becoming an ever-greater threat it is unimaginable that the UK could run out of baked beans. Yet we are entirely dependent on imports. Because of the vagaries of the Great British Climate, growing navy beans for canning has not been an option until now, but finally it looks as if things might change. The idea that the UK could become self-sufficient in baked bean production is no longer just the dream of those of us who believe we need to be growing more of our own food. Can modern approaches to plant breeding deliver us a bean fit for the job? The runes are looking good.

On a lovely sunny May afternoon I was hanging out with one of the great advocates for the UK to grow more pulses, Josiah Meldrum, co-founder and director of a wonderful British business, Hodmedod's. His company supports farmers who want to diversify their harvests by growing legumes for human consumption. Together, they decide which varieties to grow and

The Accidental Seed Heroes

Hodmedod's buys the crop, which it cleans, grades and markets. To meet an increased demand for pulses, farmers across the country are growing a great diversity for him, including peas, beans, chickpeas and lentils. There has been a lot of hype over the years suggesting a breakthrough in breeding will mean we can now grow all the beans we need to supply Princes Group, Heinz and other producers. I was keen to get Josiah's take on the realities. He sells ten different varieties of dry haricot bean, all British grown but none of them small, white and suitable for canning – at least, not yet. But he believes the future for the British baked bean is bright. He receives beans to trial himself from an exciting quartet of players who are collaborating in an inspiring example of participatory plant breeding in the UK: arable farmer Andy Ward, bean breeder Eric Holub, seed company Agrii and the processing company Princes Group are trialling new cultivars.[3] Together they are in the process of transforming the cultivation of navy beans for canning.

Professor Holub employs conventional methods in his breeding programmes. His focus has been on identifying the molecular characterisation of key traits of certain alleles (a variant form of a gene located at the same position – genetic locus – on a chromosome) that he needs to breed into a navy bean that could thrive in the British climate.* Until recently breeding and development of improved navy beans were conducted where they had first been cultivated and grew most easily – the US. These beans hated the British climate – too cold, too damp and not enough sunshine late in the summer

---

* Like many crops that I write about, the common bean (*P. vulgaris*) does not lend itself to transformation through the application of genome editing (GE), which is why conventional methodologies continue to prevail in cutting-edge breeding research.

A Future Full of Beans

to enable the crop to ripen and dry. Holub has changed all that. He started to develop Capulet in 2011. Ten years later Warwick University signed a deal with Agrii to promote this new variety's commercial production. After a challenging start – thanks to a drought in spring 2022 – Andy Ward, working closely with Professor Holub and bringing his skills as an arable farmer to the project, was able to harvest several tonnes of Capulet, a few of which were given to Princes Group to can. The rest are to be grown on over the next few years to bulk up seed sufficient for several farmers to produce at least 1,500 to 2,000 tonnes of beans a year. This is enough for just one week of processing into 13 million tins of baked beans – about 2 per cent of UK demand. That may not seem like a lot, but, by the start of the 2030s, the UK could be growing 10 per cent or more of the canning industry's needs. There is no reason why by mid-century we should not be entirely self-sufficient in baked beans.

I get the feeling from many of the people involved in the development and production of pulses in the UK that they are frustrated that the media is obsessed with baked beans and not looking at the exciting potential of other types. Capulet is not the only bean Eric Holub has developed. There are two others that could utterly transform the cultivation of plant protein in the UK and subsequently in countries with similar climates. They are Godiva, a blonde variety of haricot, and Olivia, which is black. These two cultivars have very different applications from Capulet: as dried beans, which are a key ingredient in plant-based meals. In food processing they could replace imports from countries like Mexico and southern Europe, where beans are a fundamental part of the diet. They also provide nutritious and low-carbon alternatives to meat.

Although Capulet has only been tested in Princes' research kitchens and I have yet to taste it, I am reliably informed by those who have that it is impossible to tell the difference between a can of British beans and an American one. If that is what the customer also thinks and if, in the next two or three years, this participatory programme is able to ensure improved and consistently reliable yield and price to the farmer, we could see a material improvement in the resilience of food production in the UK and with it improved soil health and reduced carbon emissions.

Is the Future Fava?

Fava beans (*Vicia faba*) are self-fertile, but the action of bumble bees visiting the flowers greatly improves pollination and without them there would be no cross-pollination or resultant increased genetic diversity. This explains something I have noticed when growing out fava beans for seed myself. I get a good flower set from populations grown under cover when I stimulate setting by giving the plants a good tap morning and evening. This is unnecessary for crops grown outside, because bees do all the work.

The importance of pollinators in the reproductive process that can result in great genetic diversity within populations under cultivation sets fava apart from other beans. Today, it has become the focus of a search for an alternative pulse for the UK baked bean industry. Josiah Meldrum kindly gave me a tin of Hodmedod's own, made using a popular winter cultivar called Tundra, which is best known as an animal feed. When I tried it at home I wasn't impressed, and I fear its days in a can are numbered.

A Future Full of Beans

The challenge for producers using fava beans are many. Primarily, it is about the canning process: Tundra's robust skins make for uneven cooking. This does not appear to be a problem for the American canning company California Gardens, the largest producer of tinned fava beans in the world, so clearly one of the routes to success involves scale. Josiah is keen to trial an old British cultivar called Maris Bead – bred more than fifty years ago at the Plant Breeding Institute (PBI) outside Cambridge – which is smaller, with a thinner skin and much better flavour than Tundra. But Maris Bead breaks up during cooking and more work is needed either to improve it by crossing it with a more robust variety that keeps its shape, or to find a different way to process it.

An alternative to Maris Bead could be a Scottish heritage variety currently undergoing trials at the James Hutton Institute (JHI) in Scotland: the focus of Dr Pete Iannetta, a man who is passionate about beans and their importance in providing all of us with plant-based protein. Enter the Henry Taylor bean, the result of conventional breeding in the 1970s by one of the most highly respected plant breeders in Scotland, from whom it takes its name. Dr Taylor's first love may well have been alpine plants, but his commitment to developing crops that could flourish under cultivation in Scotland was legendary. His bean emerged while he was working at the Scottish Crops Research Institute in Invergowrie, which subsequently became the JHI. It may now, finally, be the foundation for a revolution in fava bean breeding and cultivation. I am keen to try the next generation of baked fava beans and a collaboration between Hodmedod's and the JHI to commercialise the Henry Taylor is expected to be ready for tasting at the time of publication. Pete admitted to me that the challenges to bulking up and trialling are many and the road to success by no means guaranteed. One surprising use for the fava bean is in brewing. Fava bean beer is already being tested in Edinburgh. A tasty tipple, although – call me old-fashioned – I like my beer made with barley and hops.

The focus on fava bean breeding has generally been to support the animal feed market, with any benefits for human consumption a sideline. But traditional and ancient FVs like the small, black Ethiopian fava bean, with its high levels of flavanols and antioxidants, not only show great resistance to a number of diseases but taste good too.[4] In my travels in Ethiopia, I saw a lot of favas being sold in the market. These consisted of diverse populations – different FVs all being grown together. Sometimes the largest beans were separated out and sold at a premium for human consumption; mixes were mostly for animal feed. One of the most memorable dishes I enjoyed was simply called *ful*. Unlike *ful medames*, the national dish of Egypt where the beans are eaten whole, Ethiopian ful is a puree, richly flavoured with spices and eaten for breakfast scooped up with injera, the local flatbread.

An Unloved Subject for Breeders

Despite the size of its genome and its importance as a crop for a sustainable future of food, the fava bean is the Cinderella of the legumes when seen alongside the major pulses – peas, common or French beans, chickpeas and soya. Favas, commonly known in the UK as field beans, had been widely grown for centuries both as an overwintering and spring sown crop. At the end of World War II, they were considered of little use in the human diet, and breeders showed virtually no interest in improving them. Production throughout Europe

A Future Full of Beans

fell; the focus was on the US system of maize and soya as the main ingredients of animal feed. However, in 1973 there was a global crisis in soya bean production – alongside a greatly reduced cereal crop – and prices spiralled. This was due to a number of factors, with much of the chaos down to weather and crop failures in the then USSR. Reduced cultivation and an unexpected growth in demand caught governments by surprise. US beans became cheap because of a rapidly devalued dollar, so, to ensure a continuing supply at home, the country put a stop to exports into Europe.[5] As a result, and in response to a call to arms initiated by what is now the EU, the PBI undertook pioneering work to revive the so-called 'grain legumes': pea, fava bean and lupin.

The two greatest pioneers in fava bean breeding were the PBI's David Arthur Bond (1929–2017) and Jean Picard (1924–2016) from France's equivalent, the Institut National de la Recherche Agronomique (INRA). Bond and Picard focused on creating F1 hybrids because they showed great hybrid vigour, contributed to self-regulation of fertility in fava bean populations, and averaged 22 per cent greater yields. But because the process of selecting inbred lines to develop these new cultivars resulted in high levels of sterility in male plants and poor seed set, they were twice the price of conventional varieties. Farmers in developed economies who are able, with inputs, to get heavy crops might stomach the extra investment because there is a net gain of about 10 per cent, but in indigenous and low-input farming models there is no obvious benefit.[6] This is yet another example of why farmer-led selection and evolutionary breeding can be a more sensible way forward for many crops.

Applying the latest in genomic approaches, which involves working with the wild parents of fava beans, is not easy. One of the reasons for this is that plant scientists have not been able to identify a wild relative capable of producing fertile seeds when crossed with a domesticated bean. Another reason is that all the genetic diversity available to breeders exists only in the 200-year-old cultivated gene pool, limiting their ability to develop new cultivars. Despite this, beans from one part of the world are distinct and very different to beans from somewhere else far away, so traditional approaches are providing real solutions.[7] I have seen this myself, having come across fields of fava beans growing in unexpected places such as the foothills of the Himalayas in northern Myanmar and at altitude in Chile's Atacama Desert.

The Recalcitrant Bean

Traditional methods of growing fava beans, whether for human consumption or animal feed, are based on treating a crop as a population. This means that as the climate changes, so do the mix and diversity of beans within the population. Going down the road of homogeneity by growing hybrid beans does not provide a solution to food security because the wretched farmer must buy bean seed every season and the breeder must be alert to the biotic and abiotic changes that require them to select for traits that work in the moment. Evolutionary populations are continuously doing what the description tells us: evolving. I find that both exciting and comforting, and it's something I have seen a lot of in indigenous farming systems around the world.

It might be a blessing that, unlike soya, which was taken over by global agribusiness and became a poster child for GM, the humble fava bean was only of interest to academics and freelance breeders; there is not enough money in it to warrant the investment for profit-led programmes. Breeders are not kidding when they accuse the fava bean of being 'recalcitrant', meaning it doesn't like to play ball when subject to molecular breeding techniques.* The fact that its genome is not fully referenced means it cannot be slotted into a programme driven by the gene-editing tool CRISPR/Cas9. However, the fava bean is one of the best fixers of nitrogen of all legumes and this is the case even when there are already high levels of nitrogen in the soil. This makes it a particularly valuable crop for low-input systems when rotated with cereals. Many FVs show some resistance to devastating diseases like chocolate spot, and breeders in many countries around the world have released new varieties showing partial resistance.[8]

Another horrid disease is rust, something most of us growers of broad beans will have encountered alongside chocolate spot. A number of rust-resistant genes have been identified and some are able to be tagged so that they can be used in marker-assisted selection, which should enable breeders to 'pyramid' genes to build greater disease resistance.[9] But any assumptions that modern plant breeding is going to come up with a super-fava that will be embraced by all who grow and love this crop need to be moderated. Conventional breeding and evolutionary in situ improvements – the result of participatory breeding involving farmers – are likely to prove the most resilient methods for meeting the needs of hungry animals and people into the future.

---

* Recalcitrant has a number of meanings botanically. Recalcitrant seeds are those, like mango and avocado, that cannot be preserved conventionally by drying or freezing. In breeding terms, recalcitrant species are those that do not respond to molecular breeding approaches using tissue culture – in vitro and gene editing being the two main ones.

Because it is nutritionally dense, the fava bean has much to offer. It should be a key ingredient in agroecological methods of reducing carbon-based fertilisers and supporting pollinators. Producers of processed food are attracted to fava bean flour because it is pale and tasteless, which means it does not influence the flavour profile the producer is creating from other ingredients. Farmers would like to grow from seed that is more spherical in shape and smaller in size, making it easier to harvest mechanically, so there is plenty for commercial breeders to be working on. All that is required is for innovation and improvement to be properly resourced through collaboration between growers and publicly supported plant breeders.

So much breeding today is focused on using transgenic or mutagenic solutions as well as conventional gene editing and I wonder why they go to all the bother. These breeders and seed producers bang on about speed: 'We need to be breeding new varieties much faster with all the tools of genomics at our disposal.' But what has been achieved with this approach in the last fifty years has come at a great cost to the health of people and planet. My money is on those focused on more conventional approaches; looking for solutions – in the case of fava beans, of better adapted populations at a local level. A broad-based collaboration between plant breeders, farmers, canners, brewers and purveyors of fine bean flour can offer delicious and planet-friendly choices to hungry customers – and it is already happening.

An Example of Acting Together

Switzerland has had a part to play in my search for delicious beans. ProSpecieRara is a remarkable organisation that maintains, celebrates and supports amateur growers and

A Future Full of Beans

professionals alike to cultivate a huge diversity of heritage and heirloom crops, saving and sharing seeds. I became interested in their work as a result of my involvement with the Heritage Seed Library in the UK, which has a similar purpose. So, I was pleased to see ProSpecieRara as one of the participants in an ambitious five-year project to develop climate-resilient vegetables, intended for low-input organic production. Funded by the EU to the tune of €6 million, BRESOV (Breeding for Resilient, Efficient and Sustainable Organic Vegetable production) was completed in the spring of 2023.[10] It involved geneticists, growers, research institutes, organic seed-breeding businesses and food manufacturers from across Europe, China, North Africa and South Korea. The aim was to use their combined knowledge to exploit the genetic variations of three crops to improve productivity and strengthen the diversity of seed bred specifically for organic production: one of these was beans.

Taking a multidisciplinary approach, the participants selected pre-breeding and breeding lines that included FVs and crop wild relatives (CWRs). The ambition of BRESOV was to create germplasm to 'pump-prime the production of new seed for the organic growing sector and will also serve as a model for the enhancement of other crops'.[11] ProSpecieRara focused on dwarf and climbing French beans, using their collection as source material. By the end of the trial, working with other Swiss members of the research group, they had tested and evaluated the culinary and production potential of dozens of beans from their collection of nearly 300 varieties. With support from Switzerland's Federal Office for Agriculture and a Swiss seed producer, trials involving growing and evaluating crops of drying beans in private gardens have been conducted to see how commercially viable many FVs and

The Accidental Seed Heroes

heritage varieties can be. From a first selection of twenty-four varieties of both beans eaten fresh and shelled, one stood out as a shelling bean – a heritage variety of dwarf or bush bean called Brown Dutch. Others were seen as having potential as dual purpose. What the research also revealed is that flavour varies very little between varieties – if at all, in most cases.

What I find inspiring about this project is that it lays the foundations for a truly collaborative approach to developing cultivars fit for a changing climate, specifically to flourish in low-input, agroecological farming systems and without being subject to patent. The fact that most beans taste the same underlines the importance of celebrating their place in local food culture. As long as they are perceived as delicious by those who love them, diversity will rule!

Farmers' Variety Lookalikes

This chapter began with an homage to a big, white butter bean that is loved by the nations who consider it a core part of their food culture. The same goes for another species, which is known for its huge diversity of shape and colour: the common bean (*Phaseolus vulgaris*). In the Global South, the majority of common beans are FVs. In my quest to understand more about the diversity of approaches to breeding I have spent many happy hours wandering in markets around the world. Indigenous crops are the bedrock of sustainable and adaptive solutions to feeding ourselves. But what I find equally inspiring is how we have strengthened food cultures by integrating new foods, and at the same time built more resilience into local economies. This was very evident in Ethiopia.

Phaseolus vulgaris arrived in East Africa from the Americas early in the sixteenth century and speedily integrated

110

into indigenous farming strategies. The result today is that the country is home to a considerable diversity of FVs that fall into two distinct types: deep red and black for drying, and speckled as dual purpose. Naturally, farmers were selecting and maintaining for resilience and, like their counterparts in South America, would grow different varieties together. The result today is fields of mixed crops of locally adapted varieties, now also including some Ethiopian-bred and imported cultivars.

The Ethiopian government has been involved in improving bean varieties and importing high-yielding, elite cultivars that subsistence farmers will often willingly grow because it gives them surpluses they can sell. But alongside the new cultivars from which they might save seed for a year or two before buying again, they continue to cultivate their FVs because it is these that are the most resilient, even if they don't give as good a yield. There was no shortage of beans for sale in the markets I visited and they were almost exclusively improved varieties. To my untrained eye they looked no different to the local varieties that could be found occasionally, usually sold by women who had bought directly from farmers or middlemen.

Indigenous farmers have embraced these and other introduced New World crops enthusiastically, but not all introductions go according to plan. The Ethiopian government is intent on increasing food exports and this includes pulses – it exports navy beans to the UK, for example. It has also recently decided to encourage farmers to grow mung beans (*Vigna radiata*). These are a useful agroecological crop because they are excellent nitrogen fixers, with deep roots and plenty of biomass. They are native to the Indian subcontinent and were domesticated at least 3,500 years ago.[12] On the face of it, this bean might not seem well suited to

Ethiopia, as it does not tolerate drought or heat stress and is also disease-prone, but it can grow at elevation.[13] The Ethiopian government saw potential for this fast-growing, high-value crop, known locally as *mascho*, for export to India and Canada.

Cultivation got under way in 2019 and the few farmers in the south who grew it were very happy because they got two crops a year and a good price for the harvests. But then more farmers wanted to grow mung beans, including in the north. Although they were encouraged to save seeds for ongoing cultivation, yields began to drop and fresh seed was expensive. After a couple of years farmers could barely break even; at worst they lost money. Today, mung bean is a niche crop grown only in the south, where it is still possible to make a profit because fewer farmers are now growing it. Some find the crop adds another layer of diversity and resilience to their farming because animals can graze the young plants a couple of times before they flower and set seed without it affecting yield.[14] Because it is a leafy plant it makes great hay, too. It will be interesting to see how, over time, farmers integrate this new introduction, both for food and for fodder.

Inspiring Us All to Eat More Beans

The familiar but boring trope that beans make you fart is not one I shall be repeating here. Needless to say, research shows that the more we eat legumes in all their forms, including beans, the less we fart[15] and I aspire to emulate most Spaniards in eating some form of pulse every day. My pursuit of deliciousness in one of the most important food crops for our health brings me into contact with many people for whom the bean is a true obsession; none more so than members of

A Future Full of Beans

The Global Bean Project. To sit in on their regular webinars, to talk with them about what constitutes deliciousness, is to be in a world literally full of beans. You will find members in at least forty countries around the world, although mostly in Europe; all are passionate about promoting and celebrating beans as the bedrock of many food cultures, for their diversity and their importance in ensuring a sustainable and healthy future for us all.

Beans are the go-to plant protein and are making inroads into meat consumption — albeit modestly. But there is increased momentum and, as we have seen, a broad front of breeding approaches. Locally adapted bean species, stalwarts of food cultures everywhere, deliver a powerful message of inspiration. As well as providing greater food security, they have a significant and positive impact not only on human health but on that of our soil. I believe the stories that give life and personality to beans at home and abroad empower you, dear reader, to be part of the solution: one filled with hope and pleasure. To feast on the bounty of plant diversity that nourishes us all is to be truly alive.

CHAPTER SIX

Cultivating Capsicums
A Tasty Future

*The idea of the species, in short, rests upon
the fact that all the individuals of which it is
composed are, to an indefinite extent, capable
of being fertilized by one another and only by
one another.*

M.M. Vilmorin-Andrieux,
The Vegetable Garden (1885)

It was one of those delicious Balkan afternoons in early September, more than warm enough for T-shirts and naff shorts. I found myself wandering beneath a deep blue, early autumn sky with the lowering sun sending spears of light into dense rows of peppers. I was following in the footsteps of a master grower and breeder, Astrit Kadilli, who lovingly shepherds his flock of regional FVs of pretty much everything he grows on the hectare of land that surrounds his house on the outskirts of Korçë. The object of our attention: a pepper the size of a beefburger and infinitely more delicious. I was back in one of the loveliest and least visited corners of Europe, the southeast of Albania, among the mosaic of market gardens and small farms that assure this region of its unique place as a crucible of European crop diversity.

Cultivating Capsicums

I had wandered among the fields outside the surrounding villages earlier in the year to witness farmers planting out their vegetables, including a favourite pepper. Now, with crops on the long narrow strips of field approaching the end of harvest, it was time to savour for myself one particular specimen: a deep red beauty whose reputation as something truly delicious was never questioned by anyone I met who loved growing and eating it.

The vegetable in question was a local FV sweet pepper called *Gogozhare e Madhe* (Big Gogozahare) – one of two similarly shaped but different-sized varieties, the other being, unsurprisingly, *Gogozhare e Vogël* (Little Gogozhare). I admired these 5-inch (12 centimetre) diameter fruit that looked as if they had been run over by an absent-minded tractor driver: they glowed deep crimson within the shadowland of the plant's interior. Pretty ugly might be a fair description, but when I pressed my thumbs into the flesh on either side of the stalk to prise it open as I would an oyster, and bit into the meaty flesh, I experienced both taste and texture that would put any meat patty to shame. A mix of savoury fruitiness, yet displaying a soupçon of sweetness, Astrit's pepper presented me with serious culinary competition to my favourite pepper – the much-loved Ukrainian one I had discovered in Donetsk in the late 1980s, which had set me on the path of being a seed detective.

What's in a Name?

A vegetable's significance and place in local food culture are key to maintaining flavour, diversity and resilience in the face of climate and commercial pressures. In Albania, Gogozhare is more than just a name, it's a story in itself. A description of this type of pepper associates the name Kokzhare with a variety known as Stambolle – in translation, 'Istanbul pepper'. It is common in Albanian to replace the letter 'k' with 'g', so Gogozhare also translates as 'Istanbul pepper'. The annual Kokhzar fair, which began in 1867 and continues unchanged to this day, takes place in the village of Oyil in western Kazakhstan, once an important stop on the Great Silk Road. Dr Robert Damo, who heads the agricultural faculty at the local university in Korçë, suggested that the Gogozhare pepper is connected to this fair; either introduced directly to Albania by merchants or via the great markets of Istanbul late in the nineteenth century.[1] Which brings me back to the pepper I found in Donetsk, Ukraine, in 1989 and the diversity of FVs we see flourishing around the world.

Gogozhare and the Donetsk pepper have many similarities. Multi-lobed and thick-fleshed with complementary organoleptic properties, these two FVs – one from the southern Balkans, the other from southeastern Ukraine – are cousins, certainly, with a common immigrant ancestor that arrived in the eastern Mediterranean in the early years of the sixteenth century. Sweet and gently spiced peppers were embraced by the Ottoman Empire in those days and became a fundamental ingredient in local cuisine. They are just two examples among the great diversity of multi-lobed sweet peppers from central Europe that show how appearance alone belies the fact that they are different versions of the same variety. Their morphological differences are the result of a process of continually adapting through farmer selection to the various pressures they find themselves under today. This is why it is so important to nurture and celebrate them. The pepper farmers of Korçë are on the front line in the conservation of this genetic diversity and their skills in selecting

Cultivating Capsicums

seed of the most resilient as well as the most delicious new generations of their fabulous pepper should be encouraged, supported and emulated.

Meeting Gogozhare, savouring its glorious flavours and enjoying it in its various forms in the local restaurants brought home to me the power and importance farmers across eastern, central and southern Europe have had in selecting from the countless mutations and accidental crossings that they will have spied among their crops. The result: a veritable cornucopia of shapes, colours and flavours to suit both the taste buds of their compatriots and also, crucially, the climate, soil and rainfall of the places in which they have come to flourish.

Korçë is on a plateau at an elevation of about 2,800 feet (850 metres). Its climate, although continental, is very different to that of Donetsk, which is only a few metres above sea level. Both regions, however, enjoy long, hot summers and have a rich agricultural heritage. The two peppers are equally important parts of their respective food cultures. The challenge facing this most wonderful of foods is how to ensure it can continue to adapt and evolve with climate change and retain all the qualities so highly valued by those who eat it. I cannot emphasise enough the importance of championing, nurturing and helping FVs on their adaptive journey through time. And this is something Astrit is a past master at.

You Cannot Have Too Much of a Good Thing

Gogozhare was not the only local pepper Astrit was growing. His approach to seed saving felt like anathema to me, but there was method in what I first perceived as madness. In what

The Accidental Seed Heroes

looked like random planting, Astrit had mixed among the Gogozhare different-shaped peppers – longer, yellow and red ones – which he called *kapi*. Ones that he liked to harvest green he called *kapi Egjelbër*, after their colour. Like all the farmers I met that autumn, he was tending his collection of dozens of local varieties of fruits and vegetables – all FVs, or traditional Albanian breeds – through careful selection, saving seeds of the best examples of each type. Because he had different varieties growing in close proximity, there would be accidental crossing and the occasional mutation. Having savoured these 'accidents', when he tried one that he felt warranted tender loving care he would include it in a lengthy process of selection, one that could take many years. Hence the rich diversity of shapes, colours and sizes, all flourishing on his land.

Farmers are the best company for a vegetable anorak like me and spending a sunny afternoon with Astrit admiring his crops reminded me of how anyone growing, saving and selecting seeds over many seasons is contributing to building more resilience into our food supply. In the thirty-five years or so that I have been growing the Donetsk pepper, I have saved and selected seed from the first and largest fruit every time I have grown it – over twenty times. That's twenty generations. Today, my pepper is the first to ripen in my greenhouse and is happy growing outside. Now I think it is fair for me to call it a Welsh FV – possibly the first locally adapted pepper suitable for outdoor cultivation in the UK. It can be held in the Heritage Seed Library to be maintained and shared among members long after I have shuffled off this mortal coil. It is now being employed as a genetic resource for the next generation of open-source plant breeders who, using traditional methods of cross-pollinating and selection, can make seed of this pepper available to far more people than

Cultivating Capsicums

I can and create new varieties that will flourish not only in the UK's changing climate but maybe in other countries too.

A Welsh Accident

I grow lots of capsicums, from super-hot varieties of the species *Capsicum chinense* to the fruity and less powerful *C. baccatum*, the decorative, colourful *C. frutescens* and the many types of *C. annuum* that cover the full panoply from mild and sweet peppers to hot-as-hell chillies. Although capsicums are self-fertile, they can readily cross, including between species, so to ensure seed remains true to type it is necessary to isolate one type and variety from another. But accidents happen and that is what makes the life of an accidental plant breeder like myself such fun!

There is a lovely mild French heritage chilli called *Doux des Landes* (Sweet of Landes), and I have grown it and shared seeds for a number of years. While writing this book I received a letter from Cambridge academic Martin van Rongen, a fellow lover of capsicums to whom I had sent seed. He told me that he had grown fourteen plants from fourteen seeds and they were all a bit different and rather hotter than expected (although there is some heat in this chilli at the best of times). This was clearly an example of accidental crossing, so now I must go back into my notes to see what else I was growing nearby and how my isolation might have failed. However, even if I cannot find clear evidence of accidental crossing – after all, the seed I was given may already not have been true to type – perhaps what I now have in my collection is a new Welsh variety I will call *Melys y Cymru* (Sweet of Wales).

Not only are capsicums of all types prone to crossing and easy to work with, they are heterogeneous and therefore

119

The Accidental Seed Heroes

capable of producing mutations and new forms with great regularity. Of those fourteen 'lines' of Doux des Landes, Martin identified some variability in heat from hot to very hot. He told me that the peppers, when dried, make the most delicious chilli flakes, which he eats every day. He hasn't expressed a desire to keep seed of his 'rogue' chillies, but when these 'accidents' happen in our own gardens it is an opportunity for us all to be plant breeders. Selecting the seeds from a particular line – as I did with my Donetsk pepper – is the starting point. Now isolation is key, because one is trying to select for consistency and specific traits. There is no guarantee of success, which is why plant breeders have to be patient and philosophical. Genetics does not follow a straight line, meaning that failure is always more common than success!

A Great British Breeder

At the time of writing, the world's hottest chilli is Pepper X, having replaced the Carolina Reaper, which were both bred by Ed Currie in the US. It tops the Scoville heat scale at 2.69 million units – a level so idiotically high that it is, for all sane purposes, inedible.* These super-hot chillies represent for me a pointless exercise in one-upmanship; what

---

* The Scoville heat scale is a measure of units (SHU) developed by an American pharmacist, Wilbur Scoville, in 1912. A precise weight of dried chilli is dissolved in alcohol to extract the components, capsaicinoids, that make chillies hot and diluted in sugar water in ever decreasing amounts, to be tested by five expert tasters until the majority can no longer detect a difference in the heat level. Dilution is rated in multiples of 100 SHU from 0 for modern hybrid sweet bell peppers to 2,693,000 for Pepper X.

Cultivating Capsicums

the world needs is a diversity of delicious capsicums that can grow in different climatic conditions and that everyone can enjoy. Currie and his fellow hot chilli-breeding specialists are part of a long tradition of mostly amateur growers selecting for extremes to win accolades and make money, be it from the hottest, the largest or the heaviest of whatever vegetable is the focus of their obsession. I digress.

Joy and Michael Michaud of Sea Spring Seeds grow and sell a hundred different varieties of chilli from seeds they have saved – many they have also bred themselves. Their business, nestled on a south-facing slope overlooking the English Channel in Dorset, is where they made chilli-breeding history by developing what was, in 2006, the world's hottest chilli: the Dorset Naga. It registered just under a million units on the Scoville scale. But it is not for developing a very hot chilli that I am interested in their story.

Agronomists by profession, for the last thirty years or so Joy and Michael have shared their passionate interest in capsicum – especially chillies – with an ever-increasing customer base in the UK and beyond who want to savour the delights of their labours. However, their approach to breeding could not be more different to that of the Albanian farmers who maintain FVs. One thing they do have in common – and with so many other independent breeders – is that they have never made a deliberate cross. They are opportunistic – dare I say accidental – breeders. What the Michauds and their peers have is a keen eye. Their selections for improvement are driven by observation.

Most years they will trial a dozen sweet peppers and more chillies. They scour seed catalogues from across Europe for open-pollinated varieties, which they are free to save seeds from because they are not subject to copyright under plant

The Accidental Seed Heroes

breeder rights (PBR) and have not been trademarked or patented. People send them seeds too and, like me, they have perused markets for interesting-looking varieties; some come from gene banks. One example that exemplifies their approach is the very photogenic US heritage New Mix Twilight, which had been held by New Mexico State University. This very old variety was in a poor condition – it was not behaving true to type. The Michauds took several years, diligently selecting from fruits that conformed to the phenotype of the variety until it was, once again, stable and suitable for sale.

Often seeds that were once widely available are removed from the official lists of commercial cultivars and this presents an opportunity for the Michauds to recover and breed them once again. Joy and I talked about one of our favourites, Hungarian Hot Wax, which is not only delicious and versatile, but one of their most popular. Like the peppers I enjoyed in Albania, Hungarian Hot Wax exists both as a specific and stable variety and as an FV widely grown not only in Hungary but throughout eastern Europe. I grow an FV that was given me by my son Jake, a plant ecologist, who found it on a market stall while at a conference many years ago. Now this lovely variety is back in circulation and, best of all for growers in the UK, the Michauds' selection is locally adapted; with every passing new generation of seeds selected, the variety continues to adapt to our changing climate.

Because peppers can mutate with relative ease, from time to time the Michauds discover 'off-types' – chillies and peppers that show clear morphological differences. If the fruits taste good or show traits Joy and Michael know their customers want – compactness, earliness and yield – they remove all the flowers and fruits from the plant and grow it

Cultivating Capsicums

on in isolation to be sure there can be no cross-pollination with the new flowers. From that one plant they will grow three or four 'lines' the next year, selecting the best one to grow on again. This requires patience, because what they are looking for is stability. Although the Michauds' seeds do not need to go through a registration process to determine they are distinctive, uniform and stable, it is fundamental to their business model that what they say is on the packet is in it. So, they bring scientific rigour to their work to ensure their selections are indeed DUS.

The most important part of the breeding process is to scrap the lines that are no good. The vast majority never make it into their seed catalogue. But every plant holds a genetic promise, so Joy and Michael have fridges full of lines they might one day return to. As traditional as their approach to plant breeding is, it is only successful because they keep copious and detailed information about everything they are working on and have done over the decades. The field behind their house is a grid of over twenty small greenhouses in which they grow plants in isolation, employing screening techniques I too use (which is somewhat reassuring for this blatant amateur). Several large polytunnels are home to the thousands of plants they sell to fellow chilli lovers.

An Addition to Welsh Diversity

Modern capsicum breeding is focused on developing F1 hybrids that ensure that the breeder can protect their IP and see a return on investment that justifies the time and cost of this lengthy process. As with all modern approaches to plant breeding, the key driver is yield. Like pretty much all the crops we see in our supermarkets, they conform to DUS

The Accidental Seed Heroes

rules, so capsicums, whether hot chillies or bland sweet peppers, are homogeneous – uniform in shape, size and colour. But – and this is the saving grace for this family of crops – the cultural and taste qualities of chillies in particular have incredibly strong associations with different types of cuisine, which is why recipes will often name the type of chilli that is best for that dish: Judion, Aleppo, Pimentón, Mathania, Scotch bonnet, Jalapeno and Habanero, to name just a few.

This is good news, because knowing what we are eating helps connect us to not only our own food culture but to others, too. So the insidious march of the modern hybrid pepper is slowed by a universal love for open-pollinated heterogeneous varieties. Once again, I see examples of how conventional and accidental breeding builds resilience, because the new and improved varieties are ever better adapted to the changing climate, as is so well illustrated by my own Donetsk pepper.

Joy and Michael know all 150 varieties in their catalogue because they taste test every one. By Joy's own admission, not everything they sell is a culinary knock-out; but those that don't tickle the taste buds so sweetly have other qualities that people value. They make beautiful houseplants with a kaleidoscope of different-coloured fruits of every imaginable shape and size. So, her favourite of the moment changes with every season. When I asked which was her current favourite, she told me about Cardiff Queen. A new Welsh variety? Yes, actually, and the result of several years of selection from an accidental mutation given to her by a grower from the city after which her chilli is named. And what is so fabulous about it? It is a lovely shrubby plant with bright red thumb-sized conical fruit that crop over a long season. Yummy, naturally! This is what makes Joy and Michael's work in Dorset so inspiring, and I can add her recent creation to the list of Welsh varieties anyone can now grow. I hope that one day they might want to trial some of the varieties I have in my collection with a view to making them more widely available to all chilli-loving gardeners.

The Motivation for Commercial Breeders

Capsicums are an important export for many countries for whom bell peppers and chillies are fundamental to their own food culture. The world's largest producer is China, accounting for about half of global production of 33.5 million tonnes, most of which is exported in the form of chilli powder. Like other major producers, China grows high-yielding varieties specially bred for its climate. The second largest producer of chillies is Turkey, with a rich culinary heritage of using capsicums in all their forms. Indonesia is also high on the list and I have a special fondness for a local variety that flourishes in the volcanic soil of the beautiful island of Sulawesi. Indonesia is a country where the use of chillies in their diverse cuisine has blossomed over centuries; but as with many other major producers, few of its capsicums end up being exported. When it comes to export, Mexico takes top spot, with most going into the US.[2]

It was in Southern California that I saw intensive production of hybrid bell peppers at scale. Hundreds of acres are planted to deliver fresh peppers across the country during the winter months. Modern hybrids, which are absolutely uniform and can be harvested in just a few passes over the fields, demonstrate to me the fragility of a food system reliant on homogeneous and genetically narrow crops. Peppers need a lot of water – approximately 5 gallons (23 litres) per

The Accidental Seed Heroes

pepper – and they are being grown in a hyper-arid region entirely dependent on water from the Colorado River and aquifers that must be replenished from what little rainfall there is. With drought having a major effect on the region, there is a real likelihood that pepper production, and that of other water-intensive crops including alfalfa, will come to a grinding halt in the not too distant future, unless California can better manage its very fragile and dwindling water supply. This is an existential threat to the US, which was once self-sufficient in food. Now cheap imports of peppers from Mexico threaten the livelihoods of farmers and the thousands of mostly immigrant workers they employ.

My heart goes out to these farmers of Southern California, many of whom have been growing some of the nation's favourite fruits and vegetables through the winter months for generations. They are doing their best to deliver what the supermarkets demand. But an agricultural system that is dependent on growing acres of identical, homogeneous bell peppers, for a market that expects to be able to buy whatever it fancies whenever it likes, is a fragile one. I have made no secret of my dislike for the modern bell pepper and would never give it room in my kitchen. When I grow traditional sweet peppers, which are genetically diverse and highly adaptive, and I see how loved local varieties are around the world, and no less so in the US, I just wish the market would embrace and support farmers to grow a far greater diversity in a far more sustainable way. Peppers are not the only crop that need to be less dependent on costly chemical inputs. With droughts and floods inevitable, these farmers in Southern California could take a leaf out of the book of indigenous and agroecological growers and start producing a more resilient suite of varieties. Sure, their yields may be lower, but they may sleep easier at night knowing that

Cultivating Capsicums

they will have a crop to harvest come rain or shine. And best of all, the customer will have something delicious to eat.

With many major producers growing principally for their home market, it has fallen on a small number of countries, including Egypt, Nigeria, Tunisia, Algeria and in Europe Spain and the Netherlands, to dominate the fresh pepper and chilli market with intensive cultivation of modern hybrids. For the most part this means that innovations in chilli breeding are of little interest to the major agribusinesses, and modern hybrids have evolved to support a handful of exporters.

Commercial breeding has been less interested in traditional methods of cross-pollination and accessing the germplasm of indigenous or locally adapted varieties, considering them to be inferior because of lower yield and presumed susceptibility to pathogens. As with so many other crops that feature in these pages, plant scientists avoid working with open-pollinated varieties, preferring to operate with known parental lines, which they feel confident can deliver high-yielding, stable hybrids that can cope better with biotic and abiotic stresses. Capsicums are genetically diverse and their heterogeneity means they respond well to both traditional and mutagenic breeding. The heterosis – hybrid vigour – associated with F1 hybrids is very effective with chillies, and plant breeders are also working on male sterility, meaning F1 hybrids that do not produce viable seeds.[3] Great for the breeder; not so great for the farmer, who must endlessly buy fresh seed whether she or he wants to or not.

Interesting Things Happening in Eden

I first visited Sulawesi, one of the largest islands of Indonesia, over thirty years ago. Its long, thin northern peninsula,

The Accidental Seed Heroes

Minahasa, consists of a string of highly active volcanoes, constantly adding mineral-rich lava to the ever-changing landscape. I thought I was in Eden, so diverse was the flora and so verdant the fields and gardens, full of every conceivable tropical fruit and spice. Geneticists and plant scientists in countries that are large producers and exporters of chillies, like Indonesia, are combining classical breeding with irradiation using gamma rays to breed open-pollinated offspring with dramatically improved traits. The application of biotechnology with tools including marker-assisted selection to speed up breeding programmes is now part of mainstream research into breeding innovation.[4] Is it the future?

The ability to deliver high yields under more extreme weather conditions, with less water and greater pest and disease resistance, is driving innovations in homogeneous breeding of capsicums as a commodity crop. Indonesia has problems shared by many other countries that grow a lot of chillies: fluctuations in yield due to farmers sowing poor-quality seed; inappropriate applications of chemical inputs; and poor crop processing, as well as issues regarding soil health and irrigation. The aim of some scientists is to improve seed quality through modern breeding, and one of the fastest ways to achieve this is to employ mutagenics using multigamma irradiation of seeds to force mutations that can then be used to develop ideal cultivars.[5] I find this research of particular interest because the principal centre where samples of local varieties for improvement were studied was on Sulawesi, where thirty years ago I discovered a lovely cayenne-type chilli that I have been growing ever since. The research is also an example of an approach to building a sustainable solution to food production generally that focuses on the improvement of local varieties for the benefit of the local economy, with all the social and cultural gains that implies.

Just how much involvement the farmers had with the research process is not clear, but to my mind applying modern breeding techniques to improve local varieties, with the economic benefits that come with it, is to be applauded. For me, the next step in this one example is to invest in farmer training so they can become better at identifying and selecting seeds from their new crops; after all, the plants are open-pollinated just like the originals and have not been bred as sterile hybrids. Sometimes indigenous farmers need help and when this applies to crops that exist below the radar of multinational agribusinesses, as I have seen in Sulawesi, it is something to feel very positive about.

Is the Future Transgenic?

Capsicums don't lend themselves particularly well to bio-technology – GM – breeding approaches. This is because as a species they have, in scientific terms, high genotype dependency and a recalcitrant nature. What does this actually mean? In short, peppers are a conservative bunch and don't like change (genotype dependency); if they were human, they could be seen as wilfully disobedient (recalcitrant) for not behaving in a way that suits the breeder. Maybe that is why I like chillies so much. Nevertheless, major plant breeders see modern hybrids as the way forward. Yet I believe that the great diversity of FVs and their place in the very heart of food cultures around the world will ensure that traditional approaches to breeding prevail. We will continue to enjoy the chillies we love without having to opt for something altogether duller on the supermarket shelf.

The Accidental Seed Heroes

When farmers have the skills and resources to operate at an evolutionary level to improve their crops, at least chillies and peppers consumed locally will retain their character. Whether it's an amateur who loves to play with the chilli's traits or a freelance breeder developing new and tasty varieties or a farmer wanting to maintain and improve their FVs, I think the regal capsicum is safe in the hands of The People!

CHAPTER SEVEN

Red Is Not the Only Colour

The Quest for Deliciousness in Every Bite

The secret of improved plant breeding, apart from scientific knowledge, is love.

Luther Burbank,
Autobiography of a Yogi (1946)

The sun was rising through the dusty haze of a city waking with a collective hangover, the result of a day of kite flying followed by fireworks and much loud music. Along with the good citizens of Jaipur I had celebrated the sun crossing the Tropic of Cancer in unforgettable style, but now it was time for me to pile into the back of a comfy van and head into the countryside. A Rajasthan winter reminds me of a dry early English summer: pleasantly warm during the day but with the threat of frosts through the clear nights. Maybe not ideal tomato-growing weather, but I was about to come face to face with a local variety that I was assured was as delicious as anything I could grow in wet South Wales.

The Accidental Seed Heroes

I was on my way to the village of Begas, an hour's drive west of Jaipur, to meet a group of organic horticulturalists. The farms in this region formed a patchwork of small fields, some in winter green with barley and mustard crops just coming into flower and others empty after the millet harvest. Dotted like giant carbuncles across the land, great stoops of straw to be used later in the year to feed cattle stood alongside a patchwork of stately neem trees (*Azadirachta indica*). My guide Narendra, himself a farmer, had organised the trip, which started outside the village seed shop, a modest concrete block that could have doubled as a single garage. Parked across the entrance was a small white pick-up, the side emblazoned with a banner of a waving and bearded yoga guru, Baba Ramdev. A champion of organic farming, his name is associated with companies promoting sustainable and mainstream alternatives to intensive chemical farming in India. So, I thought as I entered the little shop, is this the place to find locally bred organic seed and supplies? It was with mixed feelings that I surveyed the scene. Every shelf, if not stuffed with modern hybrids produced by seed giants Syngenta and Bayer, was groaning under the weight of bottles of multivitamins and traditional Ayurvedic tonics for cattle.*

A Tomato for All Seasons

A short drive beyond the village we turned down a bumpy track. On either side were stands of various vegetables,

---

* Ayurveda is a form of traditional medicine first taught by the Hindu god Dhanvantari, and has been practised for over two thousand years. It is widely used across the Indian subcontinent and greatly respected by the vast majority of devout Hindus.

Red Is Not the Only Colour

including tomatoes growing under billowing lengths of fleece. The previous night there had been a light frost and the fleece was intended to protect the ripening fruit. We were met by a striking-looking young horticulturalist, Shankar Sharma, who owned the 15-acre (6-hectare) farm with his sisters – whom I could see working in the fields with a handful of other women. They were weeding a variety of crops including turnips, broccoli, aubergine and courgettes. Everywhere marigolds were in full flower, nestled among the courgette plants and mingling with the tomatoes. A fantastic companion crop that helps to ward off aphids, the marigolds were also an important source of farm income, picked daily and sold in the local market. But it was the tomatoes that most interested me. I had to taste them.

Shankar was growing two varieties; one a standard type he trained up strings under a makeshift canopy of polythene stretched between wooden poles. This was one of two hybrid cultivars he was cultivating. The other was a medium-sized, round bush that the locals called Abhilash, which references the name of the original cultivar that had been grown in the region for many years. Like many of his neighbouring farmers, Shankar saved his own seed. Protected though the stems and fruits were by fleece, the frost had nonetheless blackened some of them. Even so, large trusses were happily ripening. Naturally, I was wondering if I could grow them at home and would the fruits ripen late into the season when temperatures were low, just as they were that morning in Rajasthan? By selecting seeds from hybrid cultivars to grow on season after season Shankar was dehybridising Abhilash, a process I explore later in this chapter. By selecting seed from fruits that best withstood

The Accidental Seed Heroes

sub-zero nights he was, accidentally or otherwise, creating his own FV.

A familiar routine then followed. First to taste. Shankar was especially fond of Abhilash and thought it most tasty and excellent when cooked in a masala. It was not especially sweet and had a thick skin, which was probably key to surviving the climate. To be honest, I thought it pretty uninspiring. But who am I to judge? Shankar was not alone in the region in considering his tomatoes utterly delicious. With his blessing, I selected three prize ripe specimens from which I could extract seed later.

I also wanted to see what research was taking place into improving tomato crops in Rajasthan, which meant a trip to the Agricultural Research Station at Bikaner in the north-west of the state, on the edge of the Thar Desert. There I met Ramesh Kumar, studying for an MSc in agronomics. A challenge facing growers all over the world is how to extend the growing season in some of the most inhospitable environments. Ramesh was interested to see if crops like tomatoes, which are fiendishly difficult to grow in the extremes of climate found in Rajasthan (temperatures range from sub-zero when I was there in February to 50°C in the height of summer), might perform better with various amounts of different-coloured shading. I was fascinated to see a dozen enclosures clad in green, red, white and black nets offering different percentages of light entry and of specific wavelengths. Ramesh was growing a popular hybrid cultivar, Vishwanath, which seemed to be especially happy with 50 per cent green shading! Hopefully, his research will show that the effects of extremes of heat and cold on tomato cultivation can be mitigated by different colours of shading to optimise yield.

Red Is Not the Only Colour

I was curious about how farmers in this region were able to manage with limited rainfall, being dependent upon tube wells (bore holes) to irrigate their crops. Over-extraction had lowered the water table and increased salinity. Finding crops that can cope with salty water is one of the great challenges in plant breeding. Narendra had given me a packet of tomato seed bred by the Indian company Kudrat Seeds, which was being grown by some other farmers in the region and which he believed showed some saline tolerance. Euphemistically called 'natural' seed, the name 'Tomato Kuber 7' didn't help me determine whether it was a hybrid cultivar or open-pollinated. The only thing to do is to grow it myself to see if the picture on the packet resembles the actual crop and discover just how tasty and useful it could prove to be. Might it too be a variety that would be suited to growing in the UK and of interest to breeders? At the time of writing, I have yet to put this variety into my growing plans but look forward to seeing how it and Abhilash do ere long.

An Austrian Odyssey

The tomato has become the focus of amateur and professional breeders determined to use classical breeding techniques to create delicious new varieties that resist killer diseases, and I have been fortunate enough to trial some. As a seed saver I like to grow a small number of additions to my tomato collection every year, sufficient to enjoy eating and to produce plenty of seeds. Needless to say, come the end of April it's decision time. How much room do I have to plant out sufficient to keep me fed through the year and what am I to do with the surplus seedlings, the result of too exuberant a sowing binge earlier? Invariably it is to the village shop that I hand my surplus. Some customers are always keen to try something new and names can prove irresistible. So, it has been for a new arrival from Austria: Awesome Emma. But where did it come from and why the name?

I had been invited to give a lecture tour in Austria and was keen to learn more about what was going on in the world of local regenerative agriculture and participatory plant breeding there. Enter Josef Obermoser, an earnest, slender cineaste from the beautiful region of Styria with a passion for breeding veggies that can flourish in Austria's continental climate. In the previous six years he had been working with fellow amateur breeder Ulli Klein, developing and perfecting a tasty tomato that would flourish outside in a low-input environment of organic horticulture. Josef and Ulli's proudest creation, Awesome Emma was the result of crossing two varieties, themselves the work of like-minded breeders. One was Stripes of Yore, a colourful creation of Tom Wagner, an American who is best known for giving the world the very popular Green Zebra tomato. Emma's other parent is Bianca, a tiny, pale-yellow cherry tomato developed by one of Europe's most prolific freelance breeders, Reinhardt Kraft.

Awesome Emma feels at home in the foothills of the Alps in southern Austria. Because it is open-pollinated, it can easily be used in ongoing breeding in other parts of the world, which is exactly what Josef wants to see happen. That's why I am so excited by a global cohort of tomato breeders doing amazing work. Josef and Ulli's tomato is just one of a new generation being bred for blight resistance that

Red Is Not the Only Colour

can be grown outside at field scale. And like many other varieties, traditional and modern, Awesome Emma will become a resource that plant breeders everywhere can work with to develop their own new, locally adapted cultivars. As I write, Josef tells me that several online seed sellers in Europe have his tomato in their catalogue already, and are selling hundreds of packets.

Unsurprisingly, when I met him he extolled his tomato's rich and complex flavour. And he was right so to do. It was with much anticipation that in mid-July I picked the first fruits from a couple of plants I had growing in a sheltered spot in the garden. I was not disappointed. The fruits, which are two-tone purple and gold, emit an inner glow as they ripen; it's as if the sun is shining from within. The fruit has a complex flavour and a wonderful balance of sweetness and acidity when harvested fully ripe. Popped into the mouth when plucked from the vine, warmed by the sun, it will put a smile on anyone's face.

Josef and Ulli are two of a growing number of open-source plant breeders I have met in my travels across Europe who are keen to share the fruits of the labour. Josef was pleased as punch when I said I would grow out Awesome Emma and share seeds with some producers and breeders I have got to know in the UK and around the world. I hope that with this tomato being cultivated in diverse locations, a new generation of offspring will emerge in the coming years, adding to the diversity and resilience of crops in my homeland and beyond. Josef and Ulli's tomato is just one of hundreds of new vegetable cultivars being bred by amateurs and professionals around the world. And is it delicious? Yes.

The Accidental Seed Heroes

Beating Blight

As well as giving lectures in Austria, I was on the trail of other tomato breeders. Josef and Ulli are passionate amateurs, but there is also a cohort of professionals on the front line in developing new and delicious strains of blight-resistant tomatoes suitable for outdoor cultivation.

It was early summer and the first sunny day for weeks made walking in the gently rolling hills around Graz a pleasure. Farmers were out in force cutting hay; all around me the countryside seemed to be in a very good mood! My destination was a small farm, Jaklhof, run by Anna Ambrosch, a woman with a passionate commitment to growing and breeding delicious vegetables, and in particular tomatoes, that would flourish in the often challenging climate of her neighbourhood.

Anna is part of a network of freelance breeders and university research departments across Austria, Germany and Switzerland who are collaborating to develop new cultivars that are resistant to late-season blight (*Phytophthora infestans*): a thoroughly nasty disease that most tomato growers will have experienced at some point in their lives – myself included.* It is a particular problem when growing tomatoes outside in the many parts of the world with relatively high humidity and rainfall. Depending on the weather, starting in early summer, infected plants show symptoms with leaves turning brown and rotting. The stems and fruits soon follow suit. The culprit is oomycete, an airborne fungus-like organism that thrives in damp, mild weather. It can decimate crops,

---

* The Bauernparadeiser-Projekt, administered by Arche Noah, champions organic horticulture in Austria. It supports many growers, including Anna Ambrosch, in participatory breeding programmes developing low-input blight-resistant tomatoes.

Red Is Not the Only Colour

including potatoes, unless treated with fungicides. That's why organic farmers and gardeners want to grow crops with natural resistance.

Over the years, major plant breeders have undertaken a lot of work on developing resistant cultivars, with varying degrees of success. Commercially bred types are available in most seed catalogues. They are almost exclusively F1 hybrids, so not suitable for conventional seed saving, although dehybridising (a concept I go into later in the chapter) and selecting for blight resistance through crossing with other open-pollinated varieties is excellent territory for freelance breeders. A favourite in the UK, introduced in 2015, is the F1 hybrid Crimson Crush. It is the result of a collaboration between the renowned British tomato breeder Simon Crawford and a PhD student at Bangor University in Wales, James Stroud. Although some sellers of Crimson Crush describe it as 100 per cent blight resistant, Crawford claims only 90 per cent, which still arguably makes it the most resistant of all hybrids to date. You need a particularly foul summer for it to become a victim, and even then infection is limited. As for taste and flavour, they are not only highly subjective but are also dependent on weather, watering, feed and the quality of husbandry employed; so, the fact that it is not a favourite of mine could be down to me and my growing conditions. However, Crimson Crush has many fans who love it. So, for those of us who grow tomatoes outside, it can be a reliable addition to the table. But as sure as eggs are eggs, we are going to need to grow more blight-resistant cultivars outside as the cost of conventional production under cover only increases.

The travails of blight in the UK are nothing compared to what faces commercial growers in central and northern Europe. The vast majority of tomatoes under cultivation around the world are grown outside. However, greenhouse tomato production is increasing at a great pace. In the US, 17 per cent of all fresh tomatoes are grown under cover, including half of those sold in supermarkets; a dramatic change, considering that just thirty years ago the amount was negligible.[1] In the sunnier climes of Spain, Portugal and North Africa, most tomatoes are grown in the open, with just 20 per cent under cover in Spain. It's a different story in the Netherlands, which grow over 600 million kilogrammes (over 1.5 billion pounds) of tomatoes under cover every year, so are not subject to the vagaries of climate that exacerbate blight infection.*

A Breakthrough in Blight Resistance

Some of the most exciting work in developing blight-resistant open-pollinated varieties is happening in Germany. In the last few years, Dr Bernd Horneburg and his team, engaged in organic plant breeding and agrobiodiversity at the University of Kassel in Hessen in the centre of the country, have developed a suite of highly blight-resistant tomatoes, including a beefsteak type called Vivagrande and three different-sized salad tomatoes: Rondobella, Resibella and the most blight-resistant of all, Primabella. These are all now marketed through a company that focuses on outdoor tomatoes and is committed to open-source plant

---

* Tomato cultivation in the Netherlands is the most intensive in the world. It all happens in 1,800 hectares (nearly 4,500 acres) of giant, high-tech greenhouses and is the greatest density of tomato production in the world; equivalent to just over 34.5 kilogrammes per square metre (70 pounds per square yard).

Red Is Not the Only Colour

breeding. So, growers and breeders are encouraged to make use of them in developing and improving the diversity of blight-resistant, open-pollinated varieties available around the world, which can be grown in climatic regions that are prone to this disease.

As I walked up the track into the heart of Anna's farm, I passed a couple of polytunnels, each the size of a football pitch, offering glimpses of young tomato plants growing within. Maintaining traditional varieties that breeders can then cross with modern cultivars to create new ones fit for our changing world is at the heart of the work of breeders like Anna: people who are committed to sharing their successes with a low-input regenerative model of horticulture. She is crossing and selecting for resistance in salad, cocktail and beef tomatoes and Primabella is one she is working with. To her, taste comes first but deliciousness has to be accompanied by disease resistance. Colour matters, too: the darker the tomato, the more excited she becomes. Black and purple skins are considered especially good for health as the colour is the result of high levels of antioxidants called anthocyanins, as well as carotenoids and flavonoids, which have been shown to have anti-inflammatory and anti-proliferative effects.[2] Dark tomatoes are also of interest to pharmaceutical industries and Anna says that a dark skin is a general sign of strength and resilience. Black tomatoes may be good for you, but I have yet to find one that tickles my taste buds. Can Anna improve both taste and blight resistance?

It was boiling hot in the one polytunnel Anna was keen to show me. It contained 2,500 plants, all in different stages of development. Among the dozens of varieties were many from different climatic regions in Austria. She was keen to

The Accidental Seed Heroes

observe how they performed and to see if the same varieties tasted different according to where they had first been grown. Once a tomato had passed the taste test, Anna was interested in their robustness and vigour. I was in the presence of a kindred spirit. I guess all gardeners put these two traits as a close joint second to flavour. She was keen to show me young plants of a new cross between Primabella and a beautiful American heirloom of unknown origin that produced a pink, purple and blue cherry tomato, much prized for its flavour as well as its looks. My first visit to Anna's farm was in late May, when I had expected to see her selections growing vigorously, producing plenty of flowers. But the spring had been cool, so everything was behind. The tomatoes she was planning to grow outside were yet to be planted up.

Happily, I was able to return later in the season to see what she had achieved. The poor summer weather had proved a great test for the new tomato and many other lines she was trialling. I spent an entertaining afternoon in late autumn taste-testing and arguing with her and others as to which ones I thought were worth selecting to grow the following year. I will be watching her progress eagerly.

Breeding for an Uncertain Future

Breeding new cultivars that show good disease resistance can be very hit-and-miss: perseverance is the key, as Anna has shown me. And although blight is the most common pathogen to affect production, it is not alone. One that is having a considerable impact on the movement of seeds around the world is a tobamovirus – Tomato Brown Rugose Fruit Virus (ToBRFV). It is so named because the fruits of infected plants show rugose (wrinkled) brown patches. The disease is far more prevalent in greenhouse crops than in those grown outside. There is no known control, and, to date, resistance has not been identified in germplasm collections, so there is much research to be done.

This virus also highlights a real challenge for the global movement of seeds, which is felt particularly keenly in the UK now that it is no longer a part of the EU. DEFRA takes a zero-tolerance approach to commercial imports and any indication of infection, even in just one small part of a consignment, leads to the automatic destruction of the whole lot. Plant breeders are subject to rigorous inspection of their crops, which is no bad thing. But this is having a major impact on the ability of growers in my country to source seeds, especially organic ones. So, there is a real opportunity for local breeders to fill the vacuum with a focus on both deliciousness and disease resistance.

Consumers of tomatoes want year-round supply of an ever-greater number of shapes, sizes and colours, which is good news for greenhouse growers, and this trend, the globalisation of production to meet constant demand, means that many more countries are growing ever more crops under cover. It's terrific that the world wants to eat more tomatoes – 2021 saw the highest production ever, a global harvest of 189 million tonnes; that's 23.5 kilogrammes a year consumed on average by every woman, man and child on this planet.[3] But here lies the rub: as we have seen, tomatoes grown under cover are protected from the vagaries of the weather and from diseases like blight, but the unit cost of production is higher, making them more expensive than field-grown crops. And of course you need more inputs, most importantly energy.[4] These costs and a dependency on oil for fertilisers, pesticides and herbicides just add to the existential crisis we are facing to be able to feed ourselves.

Dehybridising – Creating New Generations

I am not ashamed to admit that one of my favourite tomatoes is an F1 hybrid. In general hybrids offer uniformity in appearance, consistency in cropping and very occasionally great flavour, as is the case with the cherry tomato Sungold. It is also the subject of much work among freelance breeders to find an equally delicious open-pollinated alternative. My own experience of growing seed saved from Sungold is that more often than not the new generation of plants yields crops that are barely distinguishable from the parent. For this reason cultivars like Sungold lend themselves to a process of dehybridisation, as we shall see later.

Molecular breeding of tomatoes is big business, with nearly all modern cultivars being F1 hybrids. The alternative – open-source breeding undertaken by passionate enthusiasts and highly skilled independent breeders, alongside participatory programmes and the continued maintenance of FVs – is one way to ensure that all of us can be the beneficiaries of deliciousness and choice. Rather than eat unripe, out-of-season, tasteless and nutrient-poor tomatoes, I would prefer to enjoy fewer but tastier and nutrient-dense varieties that have been bred locally and are themselves ingredients in a more sustainable food system.

Everywhere I travel and hang out with growers and breeders there is a constant exchange of not only ideas and experience, but genetic material. Seeing this first hand and participating in it truly rings my bell: I make no apology! Many countries are much further down the line in developing new cultivars of

Red Is Not the Only Colour

tomatoes than the UK, as I saw when visiting Anna Ambrosch in Austria. But although we Brits may be way behind the curve, the dissemination and sharing of genetic material are now giving growers a much more rewarding choice of tomatoes to cultivate. Nowhere was this more apparent than when I was discovering and trialling another British-bred cultivar, courtesy of Fred Groom and Ronja Schlumberger's seed company Vital Seeds. Mango Lassi, so named because of the notes of mango in the fruit's flavour, is the result of several years working on dehybridising and then stabilising a popular F1 hybrid cultivar.

Dehybridising is a way of teasing out traits and qualities of hybrids to create new open-pollinated cultivars. It involves saving seed from an F1 hybrid and seeing what happens when you grow it. And that is just the start of an adventure into plant breeding that is dependent on serendipity and genetic accidents. I was curious to observe it in action.

So it was that I found myself one sunny summer's afternoon being shown around by Fred and Ronja, two passionate seed producers whose small but prolific operation is not far from Buckfast Abbey in South Devon. I was keen to see how they were growing their new creation, a cherry tomato that I found truly delicious. As we saw in chapter 1, when you save seed from an F1 hybrid the progeny can be very different, depending on the dominant traits of the original parents. What Ronja and Fred did was firstly grow twenty-five plants from the seeds they saved from just one plant of a cultivar called Sakura, which is popular with market gardeners for its heavy yields and good flavour. These second-generation seeds are known as F2s. They then carefully observed the traits and growing habits of the twenty-five plants and, most importantly, assessed all the fruits for flavour. Then, they used seed from six or seven of the most delicious to grow another twenty-five plants of a third generation – F3 – the following year. As well as having a great flavour, Ronja and Fred needed their new cultivar to be stable: in other words, to produce plants that were consistent in vigour and fruit that were consistent in size and taste. From the best and most stable six of the twenty-five F3 plants, they selected three or four fruits that miraculously now tasted a little like mangoes! The seeds from these fruits are F4 – fourth generation. As the strain now showed remarkable stability, Ronja and Fred decided to save seeds from a small number of the best fruits and make them available to their customers under the name Mango Lassi. The summer of 2024, for me at least, was not great for growing tomatoes: damp, cool and grey. But, of all the varieties I had under cultivation, Mango Lassi stood out as the top performer. If ever there was a case for growing lots of different types of tomatoes, 2024 proved it. With so many failures and disappointments, I was thankful to have had at least one truly British modern open-pollinated cultivar to enjoy.

Closer to Home

The telephone call from Cardiff came out of the blue. 'I'm Graham. Fancy growing my tomato? I gave the Prince of Wales [now King Charles III] some seed and he loves them!' That was back in 2018, and Graham had tracked me down after seeing me interviewed on TV. He called his tomato Tom Thumb and it was the first Welsh heritage variety I had come across. Then I met a remarkable cohort of seed producers and breeders calling themselves the Welsh Seed

Red Is Not the Only Colour

Hub. Among the small selection of seeds they were selling was a tomato called Gardeners' Ecstasy. It had been bred in Wales by a great tomato enthusiast, Tony Haigh – a journey that he began in 2008. Like so many new cultivars, it was the result of accidental breeding.

Tony had been growing a variety called Irish Gardener's Delight from his own home-saved seed, and noticed one plant with larger and slightly different fruit than normal. He considered this fruit to have a greatly improved flavour over its parent. He had been growing two other varieties adjacent to Irish Gardener's Delight: Latah, an early-cropping bush tomato with a sharp and lemony flavour that was developed by the University of Idaho; and Dr Carolyn, a delicious ivory-coloured cherry vine tomato, itself a variation of another yellow cherry tomato called Galina, which originated in Siberia. Gardeners' Ecstasy was most likely an accidental cross of Irish Gardener's Delight with Dr Carolyn.

From the seeds of one plant Tony grew a bed of twenty lines. Unsurprisingly, because the fruits were the result of an accidental cross, there was a lot of variability in shape and flavour. Each season over the next six years, from the twenty or so plants he grew, he selected just one line that expressed the best flavour and shape, only saving seeds from the best fruit of that one plant. By 2014, the crop was stable with almost no variation that he could see. Brilliantly adapted to growing in the often damp and cool weather of the region, the long trusses of red cherry fruit mature early and will continue fruiting well into the autumn. Blight resistant they are not, sadly, but a reminder of the much-loved English cultivar Gardener's

The Accidental Seed Heroes

Delight they are, fruiting well outside and able to ripen through dry autumns.*

So, why do I get excited by yet another cherry tomato that is much like a familiar favourite? Because, as we have seen throughout this book, breeding for local adaptability is one sure way of having something delicious to eat through the summer despite the vagaries of our climate. And accidental plant breeding has a tradition that goes back to the dawn of agriculture.

Indigenous Farmers Play Their Part

FVs are also a crucial part of the story of ensuring a diversity of locally adapted tomatoes that prove resilient in the face of climate change. Whenever I am on the road, I am keen to see what local varieties are being sold in markets. Genuine heirlooms are FVs that have proven resilient and are an important part of local cuisine. It is always thrilling when I stumble across the unexpected. So it was on another trip to Ethiopia. The little town of Dalbo was holding its weekly market. Unlike many countries, where the market kicks off at or before dawn, in Ethiopia they tend to start in the late morning because stallholders have many miles

---

* Gardener's Delight was one of the first tomatoes I grew over fifty years ago. Most tomato lovers considered its clusters of cherry-sized fruit to be the best-flavoured of all cherry tomatoes. Over the years, however, it has changed, or so those like me who remember it from the past believe. Has there been genetic drift, accidental crossing or mutations in commercial production? Maybe. Only comparing genomes and growing from seed that has been left untouched for half a century in a library or collection somewhere, and then doing a taste test, will provide an answer.

Red Is Not the Only Colour

to travel. They arrive by donkey cart or on foot, carrying with them everything they hope to sell. Taking up a large open area the size of half a dozen football pitches behind the main street, the Dalbo market made me feel that there was nothing I couldn't find there. Farmers had been herding their cattle and goats along the busy roads into town – the traffic was obliged to wait patiently while the animals, off to the butchery section, took their time dawdling to their doom. Carts piled high with 50-kilogramme sacks of sorghum, cabbages, maize, onions, wheat, fragrant mangoes and bright green- and red-skinned avocados, pulled by long-suffering donkeys, trundled along the rutted side streets to their allotted areas. Second-hand T-shirts, flip-flops, all manner of tools and car parts, as well as mounds of fruit and vegetables, appeared almost orderly in the chaotic crush of shoppers hunting for a bargain.

I was in seventh heaven, as you can imagine, ever hopeful of discovering something interesting and delicious. I was not to be disappointed. The tomato section appeared at first to be full of women selling only one type of a conspicuously uniform large, plum-shaped variety. Tomatoes are a high-value cash crop in this region because they need to be grown near to a plentiful water supply. Most farmers buy commercial seed, sometimes bred within Ethiopia but more often than not imported. Then I spied her: a tall, elegant and willowy woman accompanied by, as she was proud to tell me, her daughter, who was studying at the university in Sodo, the region's capital, to be a teacher. They were surrounded by a pile of hourglass-shaped tomatoes of a size where one could fit comfortably in the palm of my hand. There then followed a number of questions my driver, Ermias, was getting used to asking: 'Did you grow

The Accidental Seed Heroes

these tomatoes yourself? Do you save the seeds? How long have you been growing them? Have they been grown in your family for many generations? Are they delicious? Are they tastier than the other ones in the market?' The answer to all my questions was yes. The daughter was keen to let me know that she intended to keep growing and selling the family tomato and that her own daughter would be following in her footsteps!

The idea of cultivation continuing from one generation to the next because young people want to be farmers is one of the things that fills me with hope and optimism for the future of our food. Like other open-pollinated tomatoes – the result of deliberate and evolutionary breeding – the Dalbo tomato will flex and adapt, a unique local FV that I hope will continue to be enjoyed by shoppers in that town for generations to come.

CHAPTER EIGHT

Perfecting the Perfect Pea

Sometimes Uniformity Can
Be a Good Thing

How luscious lies the pea within the pod.
Emily Dickinson, 'Forbidden Fruit' (1896)

Certain pea breeders bring their sense of humour to naming a new variety; none more so than my friend Ærling Frederiksen, an enthusiastic amateur who, when I bumped into him one wet and chilly autumn morning in his native Denmark, presented me with some seeds of his favourite pea. He swore it was the most delicious the world had ever known, passed to him by his father, who brought seeds from Sweden in the late 1960s. It goes by the name *Æliger Ælings Ærgte Æde Ært* – Honest Ærling's Genuine Eating Pea. Over the years Ærling has selected seeds from the best pods that thrive in his corner of Scandinavia, continuing the important work of freelance breeders like him to maintain the genetic diversity of locally adapted varieties of all our crops. I am pleased for Ærling that he is the proud grower of what he believes to be the most perfect of peas, but I shall reserve judgement until I compare his with my number one. Ærling is continuing a long tradition of pea breeding and conservation in Scandinavia and, as we shall see, Sweden remains a centre of excellence in the field.

For the Love of a Pea

In my book *The Seed Detective,* I proclaimed a Catalan heirloom, Avi Juan, as my favourite pea. It has remained in the top spot since I stumbled across it in 2013; the latest in a short line that has included a classic pea of the nineteenth century, Champion of England, which hails from Kent; and Robinson, which originated in Scotland and is named after the man who grew it in the middle of the last century. This is not to admit that I am a fickle pea-lover but rather to emphasise that falling in love with a particular pea is, as in all matters of the heart, personal and liable to change.

An Avi Juan plant grows to over 2.5 metres and yields huge crops of large pods stuffed with unimpeachably sweet peas that I defy anyone who tastes them not to become addicted to. This pea has traits that the breeder, the grandfather of the wife of the grower Jesus Vargas who gave me the peas, was selecting almost certainly more by accident than design. A century ago, when Avi Juan came into the world, Mendelian principles in plant breeding were still a very new branch of botany. Farmers and hobby growers might have been deliberately crossing different varieties to develop new ones, but it's more likely that Jesus's grandfather-in-law was following in the noble tradition of his forebears and selecting seed from simple observation. Anyone who saves pea seeds will know that accidental crossing and mutations are rare indeed, certainly in my moist and cool Welsh climate. But in parts of the world where it is drier and temperatures

Perfecting the Perfect Pea

can be higher, accidental crossing is more common. Equally, Avi Juan could be an FV – in other words, synonymous with a local variety from the region. I was keen to learn more.

Because I love Avi Juan so much and Jesus was happy for me to conserve it in the UK, I handed some seeds to the Heritage Seed Library for them to determine if it was a unique variety; and indeed it is. This leads me to believe that it is a genuine local heirloom, and it has traits that suggest Jesus's grandfather-in-law wanted a pea he could harvest over a long period. It grows so tall I need a ladder to harvest the top pods. I leave plenty to ripen for sharing and saving. Normally once peas are fully mature the plant stops producing flowers: not the case with Avi Juan. Many a time I have harvested peas to save and to eat at the same time and kept spring-sown plants cropping into the autumn. As we shall see, this type of pea is the antithesis of the modern cultivar, but it has traits that a commercial pea breeder might want to assimilate into the latest new cultivars that are bred for machine harvesting. I have passed some seeds on to the John Innes Centre (JIC), which holds a vital collection of pea germplasm: 3,600 accessions plus over a thousand lines derived from the collection and another thousand lines of deliberately induced mutations. The JIC has been mapping the genomes of all these accessions, which enables them to work on specific aspects of each phenotype. Tools including genome and marker-assisted selection are employed by breeding centres to develop new assemblies of genes that can then be shared with breeders to take forward in new cultivars. But for the most part modern pea breeding still employs a classical approach of deliberate interbreeding between closely related individuals.

The Accidental Seed Heroes

You Cannot Please Everyone

Peas come in all shapes and sizes, and I have a particular love for varieties eaten whole, which include mangetout, also known as the snow pea. The edible pod pea was the result of selection by farmers in Southeast Asia many centuries ago in response to a cultural preference for eating fresh peas and pods together. The French introduced the mangetout from Holland in the seventeenth century and breeders, both amateur and professional, have been selecting and developing varieties with sweet and tender pods ever since. A favourite, and another pea I waxed lyrical about in *The Seed Detective*, comes from Laos and is therefore more closely connected to the earliest mangetout that were first grown in that region.

However, I find myself in endless arguments with fellow pea nuts about what constitutes a good pea. When I was visiting the Austrian breeder Anna Ambrosch one chilly spring day to admire her tomato-breeding work, we got talking about peas and a task she was collaborating on to develop a winter-hardy mangetout that could be harvested in springtime. Anna was just starting the programme by crossing two American types: a popular and early snap pea called Sugar Lace with a hardy field pea called James. Josef Obermoser, the skilled Austrian amateur breeder we met in the previous chapter, was with us. The conversation centred around what we thought constituted a good early mangetout. I mentioned Winterkefe, a tall and tasty variety being maintained by the Swiss organisation ProSpecieRara. Recently, I have been growing and selecting seeds from the very earliest Winterkefe pods in the hope that our warmer winters would mean I get to harvest it in late spring. In 2024, for the first time, I was able to pick young pods in early April and I was harvesting through to the middle

154

Perfecting the Perfect Pea

of June. I suggested Anna try to cross Winterkefe with a fast-maturing and hardy English dwarfing variety of excellent flavour called Tom Thumb. Perhaps she will end up with a truly delicious winter-hardy mangetout that is of lesser stature, but a superb flavour. Such are the hopes and anticipations of freelance breeders like Anna. With success comes more deliciousness and crops that can be eaten at a time of year that was unthinkable until now.

Josef asked if I had grown another mangetout that he rated highly, Beauregarde. This is a modern American cultivar, the result of a collaboration between Cornell University and the USDA. It's a deep purple type, although I consider it to be unstable because many pods from the crop I grew were green. Despite the hyperbole on the packet, I found it a miserable eating experience (I am not alone in that regard, by the way) and would never give it an inch of room in my garden again. Needless to say, Josef and I agreed to disagree, and Anna said she would try them as part of her breeding programme so she could come to her own conclusion. Perhaps she would be able to mix its genes with another pea and arrive at something more reliable whose deliciousness we can all agree on!

All One Big Family

I am unrepentant in my love of peas and have over forty traditional varieties in my collection, with many dating back over 150 years. They are almost all tall, because the wild parents that were first foraged and then selected by Neolithic farmers in the eastern Mediterranean and Near East at least 8,500 years ago were rambling climbers. They clung to host plants, in the same way as the classic cottage garden sweet pea *Lathyrus odoratus*, grown for its flowers, scrambles untidily up and along any nearby support in my own garden. It is a close relative of the garden pea that was known until the last few years as *Pisum sativum*. Today, thanks to recent studies of the pea's genome, it has been reclassified as *Lathyrus oleraceus*, a name it was first given nearly 250 years ago by the French botanist with the most fabulous name: Jean-Baptiste Antoine Pierre de Monet de Lamarck (1744–1829).[1]

The morphology of different pea species and subspecies is diverse, to say the least, but crudely all peas have either wrinkled (papillose) or smooth seeds and are the result of domestication of two wild species, *Pisum humile*, which can be found from the eastern Mediterranean to the Black Sea, and *P. elatus*, which grows further south, in Egypt, southern parts of Turkey and western Iran, on flat and fertile steppe. Recent genetic studies have shown that *P. elatus* is the most closely related to today's domesticated pea, and it now goes by the name *Lathyrus oleraceus* subsp. *biflorus*. Pods come in all shapes and sizes, and many colours!

In the nineteenth century, size mattered and some French breeders were obsessed with creating ever-larger and more delicious varieties: they came up with a giant mangetout that has found itself in a number of iterations. Firstly there is Carouby de Maussane, named after the town in France where it was first grown; a commercial variety that continues to be widely available today. My absolute favourite, however, is the Roi de Carouby mangetout; the result of work in the nineteenth century by the famous French seed company Vilmorin-Andrieux. Having grown both Roi de Carouby and Carouby de Maussane, I could easily tell the two apart – Roi de Carouby is not only bigger, it has a much better flavour. My relationship with this old variety has taken me back to Scandinavia – first Denmark and then Sweden – because, as with so many vegetables that are widely travelled, it exists there under another name.

When one is part of a network, more a family in truth, of growers and breeders sharing knowledge on seeds of their locally adapted varieties, being an accidental breeder becomes a matter of course. As we have seen, many crops that are grown in different countries and cultures are, in reality, FVs – different versions of the same type – which over many generations of selection have become well adapted to local weather and soil. Enter a famous example from Sweden called Train Driver.[2] When I grew it for the first time I realised that it was identical to – or perhaps better described as synonymous with – Roi de Carouby. It showed slightly different traits to the extent that it was ready to harvest a little later, probably because the seeds I was saving were themselves an FV. Now I am growing the two varieties together in order to maintain a Welsh FV of this rather special mangetout.

A Modern American Invention

Another pea that has gained huge popularity is the snap pea, more popularly known in the UK as a sugar snap. Similar to mangetout, it is an American creation that also has sweet and tender pods that are harvested when the peas inside are plump like a regular pea. It was the result of an accidental mutation discovered in a field of peas by one of the great seed heroes of American food culture, Dr Calvin Lamborn (1933–2017). A plant virologist, he worked as a breeder for Rogers Brothers Seed Company of Twin Falls, Idaho. He was tasked with developing new varieties to supply a pea and bean processing plant owned by the parent company.[3] The mutant plant grew like a regular shelling pea and had tender pods. Lamborn used it to cross with a mangetout called Mammoth Melting Sugar. He wondered if such a cross would result in larger, straight and tender pods full of sweet fat peas. Although he succeeded within just a couple of years of crossing and selecting, it was to take another eight before his creation, Sugar Snap, the first ever snap pea, was ready for the public. Initially treated with suspicion because no one knew what to do with it, Sugar Snap won a gold medal in the All American Selection competition for new cultivars in 1977 and was released commercially two years later. Today there are many imitations and Lamborn went on to breed a kaleidoscope of different-coloured sugar snaps. But his first remains my favourite and that of most people I know who grow it.

I am unashamedly excited about the fluid and serendipitous approach in developing new varieties and improving existing ones. Finding varieties that can extend the season, excite the taste buds and be grown by the hobby and market gardener is just as important as the work being undertaken at the other end of the spectrum, growing peas at field scale. Against a historical background of obsessive and ever-curious amateur breeders and farmers on a mission to create the perfect pea in all its forms, a collaboration between farmer and plant breeder has ensured we can always have a portion of peas on our plate. The story of pea breeding today falls into two distinct narratives. The first, as we have seen, is by enthusiasts, something that started in earnest nearly two hundred years ago and is still very much alive and kicking. The second is the industrialisation of pea breeding for the mass market, which has undergone a revolution in the last half-century. This has ensured that today there is no shortage of affordable frozen peas. The pea is number seven in the UK's Top 20 favourite vegetables behind, of all things, one

Perfecting the Perfect Pea

of the most tasteless items of modern fare – the cucumber!*
The UK is the largest consumer of frozen peas in Europe and 90 per cent self-sufficient. The pea also offers one of the most exciting and inspiring routes into growing plant-based protein around the world, with the potential to transform food security for some of this planet's most vulnerable people through participatory programmes. It is these three aspects of current and future pea breeding and improvement that, for me, distinguish it from all other crops.

The Arrival of a Leafless Pea

By all appearances, the crop of peas growing in a farmer's field 60 kilometres east of Helsinki was its usual self. Except that lurking among the ripening crop was one plant that was completely different to its leafy neighbours. It was discovered in 1950 by Viljo Kujala, a botanist from the University of Helsinki, who spied a plant where all the leaflets had changed shape. This was caused by a spontaneous mutation of the afila (Af) gene, which turns normal leaf production into tendrils.[4] At the time, commercial pea varieties not only had lots of leaves but they were at least a metre in height and prone to lodging – falling over. Cultivation then involved a labour-intensive harvest – the crop was cut and left in windrows to dry before being taken to a barn for threshing. It was not until the 1960s that a combine harvester was developed that could cut and thresh in a single pass. But even then, the morphology of the pea made the process slow and very susceptible to weather. There was a pressing need

---

* The result of a survey conducted by the South Korean food producer Jongga in 2022.

The Accidental Seed Heroes

among farmers and food processors for breeders to come up with a solution. The mutant that Kujala found was the same height as the rest of the crop, so he made no effort to use it to revolutionise pea breeding. The story of how this important mutation eventually led first to a revolution in the production of a dry pea, and subsequently one that could be harvested fresh and be frozen, was a slow and tortuous one.

The first introduction of a pea containing the afila mutation occurred in the former Soviet Union. In 1956, under Soviet plant breeder rights, W.K. Solovieva registered a new cultivar she called Usatyj 5.[5] It was subsequently pointed out by Dr Stig Blixt, one of the world's greatest pea breeders and the man who established the World Pea Gene Bank in Landskrona, Sweden, that Solovieva had done no more than take the mutant that Kujala had discovered in Finland six years previously and renamed it. Needless to say, Soviet farmers were not impressed with this tall and poor-yielding, useless new cultivar. Further work on the afila gene in the USSR was, as a result, put on the shelf.

Only in 1965 was the mutant gene given the name afila by the geneticist J.B. Goldberg, who reported a spontaneous mutation in a field of a traditional variety called Cuarentona in Argentina. Argentine breeders made no attempt to do anything with this mutant either and it was at this point that the legendary Brian Snoad, working at the John Innes Centre (JIC) in the UK, entered the picture. He took on Goldberg's mutant and in 1969, together with geneticist Stig Blixt, began serious breeding work. This was to mark the start of a race between breeders in the UK and Poland to come up with a pea 'grand design'.

At first, Julian Jaranowski, a geneticist in Poland, used gamma radiation to induce further mutations but with no

160

Perfecting the Perfect Pea

real success. He did come up with an afila-type mutation, which he registered in 1972 and named Wasata. Unfortunately, it grew to 6 feet 6 inches (2 metres) and did not yield well, so farmers gave it a resounding thumbs down and it is remembered as the second failure of the afila gene. But Jaranowski was not to be put off and, along with colleagues in Warsaw, started an extensive programme of crossings using Wasata. One of the Polish researchers, Mieczyslaw (Mitch) Kielpinski, complained that the folks at the JIC looked down on their Iron Curtain competitors and weren't interested in their experiments. However, he made a number of breakthroughs that were to fundamentally change the narrative and also form a clear divergence in research between his work in Poland and Snoad's in England. Kielpinski worked with crosses that focused on a recessive gene (St), which reduced the size of the leaf stipules – natural growths that appear on one or both sides of a leaf – and another gene, Fas, which caused flowering to occur only at the top of the plant. He called his work 'genetic juggling'. It resulted in him creating five new and distinct forms of pea known as ideotypes.

Meanwhile, Snoad and his team, working on the same genes, were creating similar plants. With the heat of competition driving them ever onwards in their search for the perfect pea, the two research groups decided independently to focus on different ideotypes. Kielpinski worked on a short-stemmed afila plant with regular stipules called semi-leafless, whereas Snoad's team focused on a short-stemmed afila type with virtually no stipules, known as a leafless type.

The first to the finishing post were the Poles, who in 1979 were granted a commercial production licence by the Polish Ministry of Agriculture for a cultivar they called Sum. That same year, the licence was sold to twelve EU countries. The first dry pea that could be harvested efficiently with modern farm machinery was born and by 1984 Sum was being grown around the world. At the same time, plant breeding was privatised in the UK, sounding the death knell to Snoad's work at the JIC and the closure of his department.

Although fresh green peas (garden peas) are a favourite food primarily in Europe, as we shall see, until a combination of advances in freezing and plant breeding arrived in the last century, it was the dry pea that ruled supreme. Its value as a fodder crop for cattle cannot be overestimated, but dry peas, either yellow or green, used whole and as a split pea, continue to be eaten in a variety of ways by humans – and in very large quantities. A British staple is the marrowfat pea, the UK's most popular form of dry pea, always available at your local fish and chip shop. Dried green peas are the most widely grown of all peas in the UK and 12.6 million tonnes were grown globally in 2022. Russia is the largest producer of dry peas in the world, harvesting over 3 million tonnes, almost all for export. However, a breeding revolution for the garden pea to be enjoyed in the same way as I do my freshly harvested ones was in the making.

The Future Is Frozen

Clarence Birdseye (1886–1956) from New York should be best remembered as the person who invented frozen food for the mass consumer market. He was a passionate naturalist and his entrepreneurial zeal started early in life – at the age of eleven he taught taxidermy to anyone who would pay him! As well as being a taxidermist and entrepreneur he was an inventor. Inspired by observing how the Inuit freeze-dried

Perfecting the Perfect Pea

fish to keep it fresh, he invented the fast freezer and in 1922 launched the world's first frozen food company, selling seafood. In 1930, Birds Eye was launched in the US, and eight years later Birds Eye Frosted Foods arrived in the UK. But it was not until 1946, after Clarence had sold his business to the food giant Unilever, that the ubiquitous frozen pea went on sale in the UK. Now owned by Nomad Foods, Europe's largest frozen food company and listed on the US stock exchange, Birds Eye is one of a suite of brands that includes other famous names known for their peas: Findus and Igloo. Birds Eye is also a brand with a business model that brings farmers, breeders and the pursuit of quality together.

This vast food production business is intimately involved in the evolution of breeding ever-better peas and other vegetables suitable for the frozen food market. It works with farmers and plant breeders, continuing innovations that started with Unilever eighty years ago. With the development of the semi-leafless or vining pea, specifically for freezing, Birds Eye was keen to differentiate itself from other producers of frozen peas by selling a premium product at a higher price than the competition. For peas to retain flavour they need to be frozen as soon as possible after harvest and for the company this meant establishing a production system that ensured the time from the moment it was harvested to the point it was frozen was no more than 150 minutes. The cultivars that had been developed by Snoad at the JIC in the 1970s made this achievable. Today, a total of 160,000 tonnes of garden peas are grown on 700 farms in the UK for the freezer trade.

Because the journey time from farm to freezer is so critical, peas need to be grown as close as possible to the factory and farmers need to be full participants in the business model devised by Birds Eye. With the primary focus on quality being reflected in the price – Birds Eye peas can cost three times as much as budget own-label supermarket brands – the consumer has to be convinced that Birds Eye peas are the best, and of course taste, colour and how the pea pours out of the bag are what persuade them to spend the extra money.

What Makes a Pea Sweet?

The amount of starch in a pea can have dramatic effects on its level of sweetness and flavour. This is what Agnese Brantestam, who heads Birds Eye's breeding programme in Sweden, is interested in because consumers love tender sweet peas! The type of seed determines key genetic traits; for example, smooth-seeded peas are wild types with a dominant gene (R), whereas wrinkle-seeded peas have a recessive mutation (r). All Birds Eye varieties have the recessive gene because the level of sucrose in these lines is twice as high as for the dominant one.* And as Agnese has told me, she rarely crosses smooth-seeded and wrinkle-seeded varieties unless she is trying to exploit a specific gene from a smooth-seeded pea.

Although yield is not the primary driver for Birds Eye, it really does matter. As we have seen with tomatoes and other crops, in a changing climate the biggest threat to yield is disease. To that end the plant scientists who are developing and improving the Birds Eye pea from their research centre in Sweden have been pioneers in selecting for resistance: in the first instance to downy mildew (*Peronospora viciae*). With

---

* Crosses of the two types result in four gene combinations in the chromosome: RR, Rr, rr and rR. The sweetest peas have the rr chromosome.

Perfecting the Perfect Pea

pressure from legislation they have been driven to innovate, successfully creating improvements that don't require large inputs of pesticide.

Another major pathogen is pea root rot, which is caused by the fungus *Aphanomyces euteiches* and has a major impact on both quality and yield. It's hard to eradicate because the spores persist in the soil for twenty years. This means that continual cultivation is a recipe for disaster and all garden peas grown at field scale are part of at least a six-year rotation, and often a ten-year one. While this is not long enough to eliminate the pathogen it is effective in reducing its build-up, and in any event peas will only be grown on land that has been tested to show no or low levels of infection. One way to somewhat reduce disease is liberal applications of nitrogen fertiliser, which is exactly what a farmer doesn't want to do when growing a nitrogen-fixing crop! Hence breeders focus on developing resistance in new cultivars.

The work in Sweden has been very effective in selecting against individual pathogens like downy mildew, through understanding the mode of plant defence mechanisms and heritability of these traits in peas.[6] Breeding for resistance to multiple pathogens is altogether more challenging because it can have a negative impact on yield and quality. Scientists cannot always have their cake and eat it! There are some biological means of crop protection for pea root rot using different strains of bacteria and fungi that can reduce the infection rate. These treatments, which are also used as biocontrols on a number of other crops, have been shown to be effective not only in controlling plant pathogens but also in helping the plant to access and make the most of nutrients already in the soil.[7]

Many pesticides and fungicides are known to have a negative impact on wildlife and biodiversity. Sweden was tougher on the use of chemicals in agriculture and legislated sooner than other countries, so Birds Eye breeders there had to work on disease resistance earlier than others. EU legislation took its lead from the Swedes. Downy mildew, like most pathogens, comes in different races and is, as a consequence, highly adaptable. Breeding resistance that can be maintained over many years in one variety is extremely difficult, but when achieved it means farmers can stick with the same variety for some time: most Birds Eye vining peas have been in production for twenty-five years. This is unlike GM soya, which requires continual reinvention to stay one step ahead of the pathogen.

When a farmer has a variety that works for them and sees no need to switch to another, what are the breeders at Birds Eye doing? Consistency of taste is the great challenge. Taste properties are a complex part of a pea's inheritance. Agnese keeps her eye on FVs and makes great use of her mutation collection to help her understand the process of inheritance: to see what happens when a gene is blocked, for example, and what part of the pea genome to search for desirable traits. Trialling is vital for her to understand what is happening within the genomes of her breeding lines. She doesn't yet know all the genes that determine taste and yield because these are inherited traits determined by many interacting genes scattered in different parts of the vast pea genome.

I am rather relieved to find that modern pea breeding follows a traditional path built around observation. As Agnese told me, you need a lot of wellington boots in the field! More than that, Birds Eye's business model puts the farmer front and centre and is something other plant breeders might take note of. Birds Eye buys its peas in the UK from a cooperative – The Green Pea Company – consisting of some 250 growers.

Perfecting the Perfect Pea

These farms are all within a short drive of a Birds Eye freezer plant. Every year the company agrees a price with the co-op for their peas. Farmers are not obliged to sell to them, but they receive a premium price the competition cannot match. Harvesting costs – which can be highly variable – are covered by Birds Eye and the risk on poor yield is covered by the co-op. Not every farmer will have a fabulous year, which means farmers who have had cultivation issues in any particular year can be compensated. As a result, many farmers have been supplying Birds Eye through their co-op for decades. Although their peas are grown as a monocrop, being nitrogen fixers they require little additional fertiliser and are cultivated as part of a diverse system of rotation that has agroecological benefits. I may rail against monocultures generally, but I am not being a hypocrite when I say peas are an exception.

Frozen peas are very much a European food. In America they are nowhere near as popular, being associated with prison and school dinners! However, the ubiquitous frozen pea has competition, particularly in France and Belgium, where bottled peas are much loved. My early memories of eating in France as a kid are of a plate of petits pois, either from a can or a jar: rather pale, I remember, but an exotic addition to a bland 1950s British diet. Andrew Whiting, the chief agronomist at Nomad Foods, told me that Birds Eye invented the petits pois. I consider this a rather engaging marketing story to ensure peas that are graded in the 'small' category when going through the freezing process can be sold at a premium. The Oxford English Dictionary notes that the first use of the term petits pois occurs in a letter by the Irish novelist Maria Edgeworth, written in 1820. You cannot pull the wool over the eyes of a sceptical seed detective!

The Accidental Seed Heroes

An Orphan Pea's Genes Transformed

Peas may be way up there in the veggie hit parade and the focus of innovative plant breeding by some of the world's largest food companies and agribusinesses. Yet, despite thousands of years of selection and domestication, there is one member of the pea family that has changed little over the centuries because, when eaten to excess, it can paralyse you – that is, up until now. It's the grass pea, *Lathyrus sativus*, which is grown across much of sub-Saharan Africa, the Indian subcontinent and even in many parts of southern Europe. Its pea-like flowers range from a gorgeous sky blue to pale pink and white and it grows to about 1.2 metres (4 feet). Its narrow, pointed leaves are nothing like the ones we are familiar with in our own garden peas. The grass pea's value is as an insurance crop – one that survives when others fail due to weather extremes. It can be eaten fresh – in small quantities – but dry peas need soaking in lime water to reduce toxin levels.

The grass pea is defined as an orphan: an indigenous crop that is an important and nutritious part of the local economy, but that falls under the radar of research organisations, receiving little if any attention from breeders. It is high in protein and thrives both in hyper-arid areas and tropical, high-rainfall regions. Its amazing traits mean it has considerable potential to improve food security and provide an excellent source of protein as well as be part of maintaining soil health because it fixes nitrogen. Eaten as part of a balanced diet it is perfectly safe, but when eaten in large quantities – which may be the case if it is the main source of protein in times of famine, for example – its consumption is associated with the disease neurolathyrism, which causes irreversible paralysis, principally of the lower limbs.

Perfecting the Perfect Pea

The neuroactive compound that causes neurolathyrism varies in degree of toxicity, with some varieties of grass pea having lower levels depending on the conditions under which they are grown.[8] But in order for this remarkable pea to truly shine as a food source of the future it needs to be made completely safe. And that requires the work of geneticists and cutting-edge plant breeding.[9]

Despite the fact that the grass pea can be grown at times and in places where other crops refuse, namely on marginal land and with minimal or no inputs, it is, understandably, not a staple. Yet, it surely could be. Until recently, the twenty-first-century professional plant breeder hadn't found the grass pea a sexy enough or economically rewarding enough subject for attention. But now it has become the focus of scientific research, which has the potential to transform this food source into something that is not only completely safe to eat but could also have a major impact on nutrition in Africa and other parts of the world.

Enter the genomic detectives at the John Innes Centre (JIC). To transform the usefulness of the grass pea, geneticists needed first to understand how and why it produced its horrific neurotoxin known as β-L-ODAP (ODAP). This is very important when attempting to alter the plant's genome to make it safe, because the grass pea also uses ODAP as a defence mechanism and to help with photosynthesis.[10] The people at JIC started by examining the genome of the plant, which was one of the latest to have been sequenced there. The insights they have made have resulted in a breeding breakthrough. Because they now understand the genome, a mix of gene editing and modern plant breeding methods means they can develop varieties with little or no ODAP in them. One of the countries where trials of new lines of grass pea are being undertaken is Ethiopia, testing how new crosses made with local varieties will perform.

As someone who passionately believes in democratising the beneficial outcomes of research, I find it deeply reassuring that publicly funded bodies like the JIC continue to offer solutions to how we diversify the number and types of crops we eat as part of building greater resilience in the face of climate change. Along with this resilience and improved food security we see better nutrition and economic outcomes for all. Well-fed folks are happier – and more productive too. The JIC sees all technologies, using both classical and molecular approaches, as tools and does not differentiate between them. Much of modern plant breeding is a form of precision engineering, so the question in my opinion is not should we be developing these tools, but how do we deploy them in a way that benefits both people and the planet. The potential to fundamentally secure a plentiful supply of locally adapted plant protein in the most climate-fragile parts of the world is huge. There are a couple of other orphan legumes – the cowpea or black-eyed pea *Vigna unguiculata* and the pigeon pea *Cajanus cajan*, both of which are widely cultivated in Africa and the Indian sub-continent. Neither is toxic and they deserve the same sort of attention now being given to the grass pea.

A Lasting Memory

With each curve in the road as it snaked its way up the mountain, a view of ever-increasing delight was revealed. Small fields of maize and beans, plantations of enset and khat lay across the gently rolling hills below, which stretched into the far distance, their pastel patchwork bathed in the early afternoon sunshine. It was my last day in Ethiopia and

Perfecting the Perfect Pea

I was on a quest to track down a pea I had been alerted to. Before I left home, I had asked ethnobotanist Philippa Ryan, who works at the Royal Botanic Gardens, Kew, for the location of a pea she had come across on a research trip into the Gurage Highlands of the Oromia region west of Addis Ababa. A farmer had shown her some particularly colourful dried peas he was threshing at the time. I was excited and determined to find the farmer and see what he was growing.

I had seen pigeon peas growing on the terraces of Konso in the south and tried, without success, to see grass peas under cultivation – I was in the wrong part of the country. Due to the ongoing civil war, I was also unable to travel to the north in order to find one of the Cinderellas of the pea world, the Abyssinian pea (*Pisum sativum* ssp. *abyssinicum*), known locally as *dekoko* – the little pea.* Fortunately the GPS coordinates Philippa had sent me proved correct and Ermias and I found ourselves at an elevation of 2,500 metres admiring the spectacular views from the field in front of farmer Jamal Awol's house, where he was busy building a new barn, helped by his teenage son Muhammed.

The young man was despatched to another barn at the bottom of a steep bank to bring a sample of the pea harvest for me to admire. And what a colourful collection it was. Jamal told me that he saved seeds every year and that the

---

* The Abyssinian pea is popular in some highlands of northern Ethiopia, where it is becoming rarer. It is loved by those who grow and eat it as it is highly flavoursome and early to mature; but yields are low and it is susceptible to a number of pathogens. Another 'orphan' crop, it needs further research to assess its potential for improvement as an important addition to the family of peas available in the region, but also for pea-lovers everywhere. I'd love to grow it!

crop was eaten both fresh and dried, either boiled or roasted. Roasted peas were then ground into flour. The plants were quite tall – about 1.4 metres – and grown without support. Jamal grew two distinct varieties together; one, named *Gondere*, had a purple flower and deep green seeds of various hues; the other, *Nech Atere*, had seeds coloured cream to palest pink, and white flowers. Both varieties were grown together as a population without selection and for the same reasons as other species populations: increased resilience and a greater likelihood of a crop whatever the weather.

The cultivation of peas in this way is an example of how we can all enrich the diversity of locally adapted varieties by growing populations in different regions, thus broadening their diversity as FVs. Mixing things up in our gardens by growing lots of different varieties of the same species together can be the start of an adventure into accidental plant breeding that delivers us more reliable harvests, and deliciousness too.

CHAPTER NINE

Let Us Eat Leaves

And Other Bitter Beauties

Plants do not speak, but their silence is alive with change.

May Sarton, *Plant Dreaming Deep*

There are times when, in the depth of a chill winter, I count my blessings. To wander into my vegetable garden to see a stately row of hardy kale, their deepest green leaves dusted with frost like icing sugar, awaiting my knife, is to see a dish in the making. I cut a few leaves to enliven my supper and the prospect of gastronomic pleasure. Love them or hate them, bitter vegetables like kale are good for us: we need them in our diet in winter because they help protect us against the cold, suppress hunger, control glucose release and counter chronic inflammation. Yet in many of our hardiest crops, we find that bitterness delivers not just health benefits but deliciousness too. Despite the best efforts of modern plant breeders to create blandness, freelance and passionate amateur breeders are ensuring that these wonderful flavours survive to be enjoyed by everyone.

The Accidental Seed Heroes

Some Bitter Truths

Bitterness in brassicas is all thanks to a group of compounds known as glucosinolates (GSLs). When they are hydrolysed by gut microbiota they become highly bioactive compounds, meaning they interact with our gut microbiome to assist the liver and kidneys in removing toxins.[1] Research has shown that cooked broccoli is better in this regard than raw – although however you prefer it your gut will benefit. Whether this enables stressed parents to convince their little darlings that eating yummy cooked broccoli is better for them than being tortured with the raw stuff, I am not able to pass comment. What is also true is that chewing raw brassicas like broccoli and kale releases GSLs. This gives a whole new meaning to the instruction, which I was continually reminded of as a child, 'Eat your greens.' There are other health benefits from eating bitter foods such as celeriac, radishes, turnips and beetroot: a high intake of these is more beneficial for people suffering from type two diabetes than eating the equivalent weight in modern, bland and sweet-tasting modern varieties.[2]

Bitterness has its sinister side, too. Plants need to protect themselves against insects and herbivores – and that includes us humans. So they produce bitter and unpleasant-tasting compounds to deter munching, and some of these can kill you very quickly indeed. Even the most bitter of brassicas are unlikely to cause you harm, but hemlock – a bitter-tasting member of the carrot family – was the principal ingredient in a concoction used to assassinate the Greek philosopher Socrates.[3] Unsurprisingly, humans have evolved to be very suspicious of anything with a bitter taste and specialist receptors in our mouth are there to ensure we spit out poisonous foods before swallowing and avoid the unpleasant

Let Us Eat Leaves

consequences of a failure so to do! Since we first began to domesticate crops, we have been very effective at breeding out toxins: certainly reducing them to the extent that they no longer represent a potentially fatal health hazard. We have come to recognise that not all bitterness is bad, but, for some of us at least, getting past our instincts is a taste sensation too far.

The Kings and Queens of Kale

Kale is the most important of the bitter brassicas and it comes in many forms. Belgian breeder Lieven David confesses to being rather sloppy in his practices; he breeds for pure enjoyment, curious to see what happens when he crosses one variety with another. If the result is edible, he might keep it going for a few years before moving on to a different cross. He hardly ever takes notes, so breeding details for some of his creations are easily forgotten. He is sanguine on the matter, as he shares his work widely with breeders who are rather better organised and occasionally his varieties find their way into seed catalogues. There is one he is particularly proud of that he has named *Doorlevende Boerenkool* – living kale. It is the result of crossing a traditional perennial kale (*B. oleracea* var. *ramosa*), known in the UK as Daubenton's kale, with another non-hearting type (*B. acephala*). Now available commercially in Belgium, its extreme bitterness means it is really only fit for pickling or fermenting. But like its parents it is pretty indestructible and thus can provide a nutritious dish in the darkest and coldest of winter days. Following the American tradition of soaking collard greens in vinegar, I like to steam the young leaves and then sprinkle liberally with balsamic or sherry vinegar.

The Accidental Seed Heroes

Because large commercial breeders are not interested in niche crops like kale other than as fodder, independent and amateur breeders are important contributors to the diversity and adaptability of local varieties. In Denmark, I was fortunate enough to meet two who deliberately allow all sorts of cabbages and kales to grow together and to cross in a very random and unstructured way. For Allan Clausen, who runs a successful organic market garden near Copenhagen, it is an important part of his business as he saves seeds of all the vegetables he sells on his farm. Walking through a stand of brassicas – a mixture of leafy kales, hearting red and green cabbages, and assorted cauliflower and broccoli – I was struck by Allan's approach to breeding new and very well-adapted local varieties. He allows the unharvested crop to flower and go to seed. He then randomly selects seed to grow the following year. His customers must love the result because his weekly harvest is always sold out.

Fellow Dane Bodil Gimbel loves her greens and, like Allan, she lets her kales randomly cross. She selects seed from between twenty and forty plants, carefully noting morphological changes. Over the years, she has created a diverse population of different coloured and leaf-shaped kales, with a particular emphasis on stem colour – both red and green. Her favourite she has called *Høj Amager*, in English High Amager, referring to her home village. The leaves are tender and at their best when eaten in late winter: a good balance of bitterness and sweetness, the result of crosses between Danish heirloom varieties. Bodil has noticed how much the character of these natural crosses changes and adapts over time. Dutch introductions into Denmark like Green of Copenhagen have been widely cultivated in the last twenty years or so and have their genes in her crosses. With greater diversity and, above all, deliciousness come increased self-sufficiency and the means to win over a sceptical public. However, even Bodil creates varieties she is less fond of: one she calls Bitter Kale. This just goes to show that even breeders can have too much of a good thing!

Mixing the Species

Freelance brassica breeding is also happening on a modest scale in England. Nestled in a quiet green valley in South Devon, two seed businesses work out of the same collection of barns on a regenerative farm. One, Vital Seeds, which I visited in Dehybridising – Creating New Generations on page 144, produces and breeds organic seeds that are especially adapted for low-input production. One summer's afternoon I was being given a guided tour by the owners, Ronja and Fred, who were keen to show me an accidental cross they had spied a few years previously.

They had been growing a particularly colourful variety of mizuna (*Brassica rapa* var. *niposinica*) called Beni Houshi. Mizuna is a bitter, leafy green that is native to Japan and has been grown there for centuries – as well as on the International Space Station in 2019. In the same part of a field, Ronja and Fred were also growing a different species: a kale (*B. oleracea*) called Dazzling Blue, redolent of the ubiquitous cavolo nero, which was bred in the US. Both species were in flower at the same time. On a path between where the crops had been growing, they had noticed seedlings emerging, which they then grew on to see what might have happened. There had indeed been a cross and they selected a number of lines of the 'accident' to grow on. Although *B. oleracea* and *B. rapa* are different species, they are closely related and can easily interbreed, as explained in a theory known as the Triangle of U, published in 1935 by the Japanese botanist Woo Jang-choon.[4]

In 2022, Ronja and Fred had a second-generation hybrid that showed an anticipated lack of stability: the plants were all different from each other. Now they are selecting lines from the best of each new generation to create stability and with it a new British hybrid, which will hopefully provide a bountiful harvest of tasty leaves for salads and stir-fries. But we will have to wait several years for them to finally arrive at a selection of a Devon-bred bitter brassica that can be grown by their loyal customers.

Next door to Vital Seeds is Incredible Vegetables, run by Mandy Barber, who is passionate about perennials. She grows hundreds of different types of fruits and vegetables that she believes are an important contribution to the diversity of foods and also offer agroecological benefits by avoiding soil disturbance, being left to grow through many seasons in one place. She, like Lieven David in Belgium, grows the Daubenton perennial kale and one other I have grown on and off for many years, the indestructible Taunton Deane. Perennial kales offer exciting breeding opportunities because they can be propagated both vegetatively from cuttings and from seed.

Like Ronja and Fred, Mandy is keen to apply systematic approaches to breeding. One brassica she has been experimenting with is Nine Star broccoli, which commercial breeders do not consider a worthwhile crop. Believed to be the result of an accidental sport* found in a row of broccoli

---

* A sport is an accidental genetic mutation that is the result of a faulty chromosome replication expressed as an observable new trait. This can be passed on to future generations.

Let Us Eat Leaves

by Mr W. Crisp of Colchester, Essex, and first mentioned in *The Gardener's Magazine* in 1907, Nine Star is in trouble because it is suffering from inbreeding depression.[5] This happens when insufficient populations of a brassica are grown together for seed, resulting in a lack of genetic diversity that leads to reduced viability.[6] Looking more like a collection of mini cauliflowers growing on a single plant and with a much stronger flavour, Nine Star broccoli is an example of a variety that is in desperate need of improving. With only a few small companies saving seeds from just a few plants, it is necessary to introduce as much fresh genetic diversity into production as possible. To that end, Mandy asks anyone who has saved their own seed to donate some that she can then mix with seed from other sources to refresh the gene pool. Now she is working with a small number of breeders around the world, and over the next few years hopes to conduct field trials to develop vigorous crops that will have long-lasting perennial habits. This requires dedicated work and lots of volunteers to tend to the hundreds of plants needed, to rogue out individuals that are off-type and to harvest large quantities of seed! Given time, more of us will be able to enjoy this wonderful – with tasty levels of bitterness – perennial vegetable, its genetic diversity restored, making it an interesting variety for breeders to develop further.

All brassicas share a distinct and somewhat sulphurous taste. This is down to an important protein that makes a number of metabolites, which are responsible for bitterness, accessible to our gut. The degree varies within species and varieties, but even relatively bland brassicas like calabrese contain sufficient to confer considerable health benefits. In Denmark, Bodil has a unique red cabbage that she calls Bodil's Kohl. Other crops

The Accidental Seed Heroes

that she has bred through random crossing include carrots and beetroot, both vegetables that contain healthy bitterness in their taste profile. She shares seeds of these valuable varieties widely among fellow seed savers and organic growers, and they will continue to be part of the grass-roots evolutionary breeding now emerging in her country.

Breeding Out Bitterness

The problem for our health is that we have, for the most part, abandoned bitterness for bland. Why? A move to ultra-processed food has fundamentally removed us from consuming a diversity of plants – bitter ones especially. Our natural instinct to be suspicious of bitter foods, an essential trait when we were hunter-gatherers, has made it easy for the main food producers to exploit this to convince us that bland is best. I am old enough to remember the eye-watering bitterness of a grapefruit requiring lashings of sugar to make it edible – for me, at least. Brussels sprouts' unappetising bitterness made them a miserable addition to the dining table on Christmas Day. Today, if you want to enjoy the true taste of traditional sprouts you need to grow them yourself or know someone who does, because modern cultivars have had the bitterness bred out of them. There are bitter leaves I loved when I was growing up: watercress, an essential part of an egg and cress sandwich until the peppery leaf was replaced with plain cress, a bland modern alternative grown in a tub on the windowsill. Watercress – grown in clear streams in southern England – is possibly the last truly bitter leaf one can still find in a bag in the supermarket from time to time.

But as our understanding of the importance of the chemicals in plants that provide bitterness increases, more of us are choosing to eat them. There is a renaissance in cultivating and breeding a great diversity of crops, bitter in varying degrees, by enthusiastic amateurs and traditional breeders. Molecular breeding is also being employed to create cultivars without the bitter taste but with the same health benefits – the horticultural equivalent of having your cake and eating it. So, is the future healthy but bland? I hope not: how boring is that?

Spoiled for Choice

I like to harvest something from my garden every day of the year and it is the winter months that offer me the greatest diversity of vegetables. I turn most frequently to salad crops that provide me with the colour, flavour and varying degrees of bitterness that are synonymous with winter veg. Members of the chicory (*Cichorium*) genus take centre stage and, for the most part, these are varieties with a long, distinguished history and names that make my mouth water. Endive (*C. endivia*) has been bred in a variety of leaf shapes: frisée (*C. endivia* var. *crispum*) is narrow and curly, whereas escarole (*C. endivia* var. *latifolia*) is broad-leaved. These types commonly have their centres covered – blanched – to exclude light for a few days or weeks before harvesting. This has the effect of not only making them look even more beautiful but of ensuring a better balance of bitterness and sweetness.

Another diverse and ecologically important species, *C. intybus*, gives us two more delicious types: puntarelle, also known as *cicoria di Catalogna*, and asparagus chicory, *cicoria asparago*. I love having these in my garden as they are the most bitter of cultivated chicory. I blanch them by tying the outer leaves together so that light is partially excluded; within a couple of weeks or so I am able to harvest dreamy, mouth-watering leaves. There are two other types of chicory that I grow every year. One is witloof (*C. intybus* var. *foliosum*), which I dig up in late autumn and then plant in a large bucket from which light is excluded to give me the classic chicory so familiar these days in supermarkets. Bitter it is not, but lovely in a salad it certainly is. The other and best is radicchio. Once caught with a little frost, its red and variegated leaves are a delicious, visual feast. The name gives away the home of this wonderful food: Italy.

Keeping Diversity Alive

Radicchio has a long history of cultivation in northeastern Italy, dating back to the early 1400s.[7] The choices we enjoy today are the result of farmer selection and breeding from one particular variety, Rosso di Treviso Tardivo (Late Red of Treviso). It has been much loved since it was first cultivated five hundred years ago and is treated in much the same way as witloof chicory – forced in darkness to give us a red-leaved 'chicon'. Over the centuries, farmers repeatedly selected and crossed individual plants within field populations to create the familiar large, round, solid-hearted plant I love so much. I am not alone in relishing radicchio, which is gaining in popularity around the world as a perennial element in mixed-leaf salads. Radicchio di Chioggia, which has a diverse and delicious number of varieties, is a global crop these days and in its homeland, the Veneto region of northern Italy, over a quarter of a million tonnes is grown every year. There remains a tradition of seed saving among many farmers such that almost 40 per cent of the crop is grown from locally produced seed.[8] Modern varieties developed from Rosso di Treviso in the 1960s include the very hardy Rosso di Verona, which crops right through the winter months and is the most successful commercial variety to date.

Accidental and deliberate crossing between traditional radicchios and other chicory species has resulted in variegated types: in my humble opinion, the most photogenic and delicious of all. For much of the twentieth century, different shapes and colours have been bred, from deepest red to almost white. Until recently, all breeding of this group of salad crops was conventional, with regional variations much loved and celebrated. But, as our demand to eat whatever we fancy whenever we fancy has become the norm, major plant breeders are starting to take an interest. The public wants their endive and radicchio to be grown out of season – in summer! There is also a need to improve existing varieties with a more systematic approach on-farm: the focus is on selection for crops able to cope with greater weather extremes, including warmer winters. Yet another example of the importance of participatory and evolutionary breeding in ensuring we have delicious harvests in the coming decades.

I remember dining on mixed salad leaves on a sheltered and sunny restaurant terrace in Florence one New Year's Day: heavenly and impossible to replicate at home. The Italians are rightly proud of their family of radicchios and because those already mentioned, plus a variegated type, Variegata di Castelfranco, are grown in such a localised region, they were given Protected Geographic Indication (PGI) status towards the end of the twentieth century. This classification has ensured that at Christmastime Rosso di Treviso Tardivo commands a price two or three times that of lesser varieties. Because forced radicchio is the result of a time-consuming and complicated process, it has never attracted the interest of major plant breeders. Molecular breeding of all types of chicory is an uphill struggle, so it is down to farmers and breeders to maintain the variety. Chicories, their flowers blazing blue in the fields in the summer sunshine, have been grown over many generations and maintained through mass selection every year; this also ensures that these culturally important – and delicious – crops continue to adapt as the climate of the region changes.[9]

More Than Just a Delicious Winter Vegetable

As well as their wonderful organoleptic qualities, chicory leaves are rich in carbohydrates, calcium, magnesium, iron and several B vitamins; so they are extremely good for us. They have other qualities, too, that can be a part of a better future for humans, other animals and the planet. The root is used as a coffee substitute – a filthy alternative to the real thing in my opinion. In parts of India, though, it is made into chewing gum, which might be marginally more appealing. It has numerous uses as a medicine, from treating jaundice and gout to relieving acute loss of appetite and rheumatism.[10] As a fodder crop it is as nutritious as legumes and better than grass. Not only does it reduce flatulence in humans, it does the same for ruminants, which makes it an important element in reduction of the greenhouse gas methane. But it is perhaps as a soil decontaminant that chicory root offers its most valuable tool, as part of a suite of measures to secure a sustainable future for food, soil biome and carbon sequestration. Chicory has a bigger root system than many other species – anyone who grows witloof can attest to that! Studies have shown that it can absorb high concentrations of lead and thus

Let Us Eat Leaves

has real value in helping remove that toxic metal from contaminated arable land.[11]

Chicory's deep roots minimise nitrate leaching and improve drainage. The result is reduction in levels of acid and salt in the soil. Because the wild relatives of chicory have been part of natural grassland for millennia, there is real potential to bring large areas of currently uncultivatable land into production as a consequence.[12] So far, molecular breeding has not gone beyond experimenting in the lab to see how genomics could be used to compete commercially with the present, mostly farmer-led diversity of all forms of chicory. Because the major plant breeders are only interested in scale, one area they might want to focus on is land remediation and decontamination, where deliciousness will be of little importance.

A concern for all of us should be the impact of introducing transgenic cultivars into an uncontrolled environment. Because wild chicory species grow in farmers' fields – it's a common grassland weed – crossing with cultivated varieties will happen freely. This is a potential nightmare for those who see the future in transgenic or genome-edited new cultivars, because of the real likelihood of gene transfer between the wild relative and the cultivated interloper. Plant breeders who patent their output and claim the IP of individual genes can't sue Mother Nature for mixing things up. Far more troubling is what happens to the wild relatives whose genome becomes contaminated – and what of the poor plant breeder who finds exciting new accidental mutations in wild relatives that contain genes from engineered cultivars? It'll keep the lawyers busy and ecologists fretting and is one compelling reason to keep research where it is at the moment, in the lab.

The Accidental Seed Heroes

Creating a Perfect Lettuce

Back in Belgium, Lieven David has also been indulging his love of lettuce. His selections of types to cross are absolutely in the spirit of the experimental amateur, following in the footsteps of countless curious individuals who have brought rigour to their work in the last two hundred years. He described the breeding journey to me: a timeline of seven years is normal. In the 2002/3 growing seasons (overwintering lettuce need to be sown one summer and harvested the next), Lieven chose to cross Celtuce, for its wild vigour (it retains spines on its leaves, as do its wild ancestors), and a lettuce called Brunia, for its cultivated vigour. He then crossed the offspring of these two types with a commercial variety called Sherwood because he wanted his lettuce to inherit its crunchy sweetness.

Over the following six years, he selected from those individuals several lines that showed the best disease resistance, were late to bolt and had crisp leaves and, very importantly for organic cultivation, a good root system. He called his creation Babel. When the plants were in flower, he would give each one a good yank. If it came out of the ground it became chicken feed! Lieven sent all his seeds to a traditional Dutch seed seller, Vreeken. Founded in 1926, this family business specialises in growing and selling organic seeds, and they sold out of Babel. Fortunately, Dutch horticulturalist Alexander Kerbusch had bought a packet and because he was bowled over by the deliciousness he saved some seeds. Now Lieven asks anyone who has seed to share it freely.

Lieven didn't stop with Babel. Between 2003 and 2012, he created more crosses using a modern cultivar called Frillice, which is widely found in supermarket mixed-salad bags, even though it is slightly bitter: it's a cross between an iceberg and an endive. Also included in his creative crossing was a hardy traditional open-hearted romaine or cos type called Crisp Mint, chosen for its sweet and crunchy flavour. In 2009, Lieven put the seeds from all his breeding lines into a single population, which he has called Opsala. Since 2012 he has continued to select from this population for diversity and, as he says, 'good-looking heads'. The world needs more Lievens and also seed savers like Alexander Kerbusch, who, as far as Lieven knows, is the only person maintaining his creation, Babel. Guess I need to save Babel too – especially if it is as delicious as my Belgian colleagues claim!

The Heritage Seed Library (HSL), which holds the UK's national collection of heritage and heirloom seeds, is home to Mescher, the oldest known named lettuce variety, bred in Austria at least three hundred years ago. It is extremely hardy with tightly packed pretty crinkly leaves, tinged pink. It has an unimpeachably sweet flavour with bitter undertones. On a trip to Austria, I visited the Arche Noah botanical garden near Vienna, a very important home for its own seed library and an experimental centre for organic crops. I asked my friends there if they had Mescher in their collection and they gave me a blank look. It appeared that the HSL was the only source for this unique part of Austria's food culture. Now, it is being grown in trial plots to see how well it is adapted to the country's more extreme climate. It is a prime example of a variety that needs to be multiplied, with valuable genes that breeders like Lieven might want to experiment with.

More Bitter Winter Vegetables to Celebrate

The rolling countryside split by the slowly wandering Danube west of Vienna is home to a remarkable Austrian grower

The Accidental Seed Heroes

and breeder, Peter Laßnig. He likes nothing more than to experiment with creating new varieties of vegetables that his customers value. I am not alone in craving colour and variety on my plate, most especially during the long and dark winter months. For over fifteen years Peter has been working on breeding a diversity of colourful hardy winter radish. It all started with an accidental crossing between a wonderful, peppery variety, Black Turkish, and unknown others in his fields. The offspring were hugely diverse, coming in many colours from white to black with colourful flesh: purple, green, yellow. It took Peter many years to stabilise a few lines, such that he can be fairly certain now of what he will get in appearance and heat level. Although there continue to be surprises, some he likes to keep and develop further and others he feeds to his chickens.

One thing Peter's Viennese customers wanted from his colourful winter offerings was something not too spicy, and he has duly obliged. He gave me four varieties to see how they would perform in my garden. The descriptions of the unnamed little darlings were black/purple, black/red, deep purple and pink/green. I sowed a sprinkling of each in a polytunnel at the end of September and enjoyed a kaleidoscopic harvest of gently spiced and super-crunchy round and conical small radishes right through the winter. There is nothing uniform about them – it's like holding a bunch of individual personalities in my fist when I lift them.

Unfortunately, complex flavours, enhanced by degrees of bitterness, have been almost completely bred out of the modern radish. Supermarkets demand varieties that will not wilt and have a uniform skin – red, of course – and pure white flesh. The poor shoppers don't know what they are missing. The opportunities for freelance and open-source breeders like Peter to respond to local demand and create radishes and other peppery roots to suit their customers' palates are legion.

My preference is for a hot radish, and this is difficult to achieve from a summer crop unless you are growing heritage varieties. So, as we head into autumn, I look forward to rather more flavoursome fare. There was a tradition in the nineteenth century to sow radish in the late summer and lift the roots for storing into the spring because they do not like frozen ground. One of the most delicious of this type is a French variety called Pasque. With climate change I can now leave the crop in the ground all winter and lift as required. By selecting seed from the best-looking specimens that have started to flower in mid to late March I have effectively created a Pasque FV that is becoming ever-better adapted to the milder winters in my garden – and without losing its peppery, crunchy deliciousness. It's another example of how all seed savers can be accidental plant breeders and add to the diversity of FVs of our favourite garden crops.

Major Breeders Get In on the Act

Molecular breeders who want to reduce the levels of gluco-sinolates (GSLs) in all types of brassica have a big challenge. GSLs are not there just to put off predators or make us screw up our faces when we bite into a particularly bitter leaf: they help plants to adapt and survive better in many different environments. Enter the science of metabolic engineering (ME). Best understood as the way to optimise processes within a cell in order for it to increase the production of a certain substance, ME's most common application is in synthesising organisms at an industrial level: it is used in brewing, cheese making and pharmaceuticals. ME is also being employed to alter the level of GSLs in a plant such that it tastes less bitter or not bitter at all.[13] However, the real interest in modifying GSLs is not with leafy brassicas to make them bland but with crops such as oilseed rape or maize that we process into oil as a biofuel, animal feed and an almost tasteless cooking oil. Maybe I am the exception, but I prefer to use cold-pressed organic rapeseed oil in the kitchen – mine is produced a short distance away in Shropshire. It cooks at high temperatures, tastes delicious, is a fraction of the price of the next best thing (olive oil) and is far removed from the vegetable oil on offer at the supermarket.

Gathering Storm Clouds

In this chapter, the stars, so far, have been crops that are of little or no interest to agribusiness or major plant breeders, all made possible by the individual characters who breed and maintain them. Modern, molecular approaches to breeding are proving challenging, although I have no doubt that at some time in the not-too-distant future we will see a new generation of crops that are no longer bitter but claim to retain the health-giving properties associated with bitterness. I am sure the breeders will want to 'protect their IP' with patent laws that include individual genes. In the US, approaches to lettuce breeding are exposing the reality of how the hegemony of major seed corporations is affecting innovation and choice: the front line is in the Pacific Northwest.

Frank and Karen Morton have, for the last thirty years, been selling organic seeds from their small farm in Oregon's Willamette Valley, a region of global importance for seed production due to its equable climate and the talented breeders and growers living there. The Mortons have built a reputation for developing and maintaining a great diversity of salad crops. They are keen lettuce breeders and, as partners of the Open Source Seed Initiative (OSSI), have made a pledge to anyone who wishes to work with their original varieties and breeding lines:

> You have the right to use these OSSI-Pledged seeds in any way you choose. In return, you pledge not to restrict others' use of these seeds or their derivatives by patents or other means, and to include this Pledge with any transfer of these seeds or their derivatives. *

Visit their website and you can buy Karen's favourite lettuce – for the moment at least – 'Karen's Fave', an open-hearted type she describes as: 'Toothsome and crisp, each leaf is glossy and muscular from its wine-red tip to its apple-green base.'[14] I call that true love, and I want to grow it myself!

Now the Mortons' work as open-source breeders is in serious jeopardy because powerful commercial breeders are applying what are called utility patents. This means that the company controls rights not only in the seed but also in all of its traits: colour, texture, disease resistance and even the way in which it has to be grown. Not only that but the patent includes the rights to further research and all future genera-tions derived from it. This means that if the Mortons (or any other breeder, for that matter) develop a cultivar that matches any of the traits of a patented one in any way whatsoever, they risk being sued for breach of patent – even if there is no

---

* This is the OSSI's wording, which can be found on their website: www.osseeds.org.

The Accidental Seed Heroes

genetic link. One cultivar in the Mortons' catalogue is Funny Cut Mix, a cross between Fine Cut Oak, a frisée type, and Jester, one of Frank's previous creations. The result is a lettuce showing a great diversity of frilly and yummy leaf shapes, many tinged with purple. You don't know what you are going to get until you see the crop growing.

Frank shared some of his seeds with another breeder to trial and see how well they performed in a different region. Subsequently the breeder told Frank he could no longer include Funny Cut Mix in his trials because it resembled in parts a lettuce called Salanova, which was subject to a utility patent with a Dutch commercial breeder, Rijk Zwaan. Funny Cut Mix's parent had no connection with the breeding material used by the Dutch breeder, but that doesn't matter where utility patents are concerned. A climate of fear now pervades the world of the small-scale breeder. To prove that her or his lettuce has not breached patent law would be costly and require detailed genetic analysis. And the terrible thing is that there might be a common gene or set of markers in the DNA sequences of both varieties because Rijk Zwaan could have been breeding from their collection of lettuce accessions, which includes germplasm of heritage, heirloom and ex-commercial varieties that have gone out of circulation, one of which could also be a relative of one of the Mortons' breeding stock.

We are seeing companies obsessively describing every detail in a new cultivar, laying claim in such a way that it becomes almost impossible not to be in violation. Some patents extend to fifty pages of description and include not only lettuce but fruits and vegetables with generic names, such as Brilliant White Cauliflower, Pleasant Tasting Melon, Red Lettuce and Heat Tolerant Broccoli.[15]

Let Us Eat Leaves

This unethical use of patent law, which was intended to improve the environment for competition in plant breeding, is having precisely the opposite effect. Even if they had the resources to fight a case in court, there is no certainty that breeders like the Mortons would be exonerated, as indeed they should be: a desire to continue to innovate and collaborate is being stifled. Now these anti-competitive US laws are being adopted around the world. They are intended to snuff out the counter-revolution in breeding and reinforce the dominance of agribusiness. If anything leaves a bitter taste in my mouth it is the Mortons' story: it strengthens my belief in developing and maintaining more locally adapted varieties of everything we grow and eat and in steering clear of growing from seeds that are subject to draconian patent law – that's positively revolting!

CHAPTER TEN

Beautiful Brinjal

The Making of an Asian Love Affair

For doubtlesse these apples have a mischeevous
qualitie, the use whereof is utterly to be forsaken.

John Gerard,
'Of Madde Apples', *The Herball* (1597)

I changed my mind about growing aubergines as a result of a chance encounter with a variety much loved in Rajasthan. It was one of the ingredients in a cookery class I attended at the house of a remarkable lady, Meenakshi Singh. Today, we know of this member of the nightshade (Solanaceae) family, *Solanum melongena,* as mostly purple. But when the British first colonised the plant's homeland India three hundred years ago, they were introduced to a creamy white, goose-egg-sized fruit they called eggplant. Americans have stayed with this name, a legacy of their British colonial connections. In the UK and in many parts of Europe, we now call it aubergine, derived from the Catalan word *albergina*; in India, it is known as brinjal. In Rajasthan, they call the fruit *baigan,* from the Arabic *ad-badinjan,* meaning devil's egg, which itself comes from the Sanskrit name for the fruit, *vatimgana.* So, the etymology is intimately linked across continents and cultures.[1] But until I got to make *baigan masala* for myself in

Beautiful Brinjal

Mrs Singh's kitchen in the beautiful Rajasthani city of Udaipur, I considered this member of the deadly nightshade family deeply disappointing. Perhaps it was eating portions of cheap moussaka in dodgy tavernas on Greek islands in the 1970s that set my heart against it. In truth, I found the aubergine a capricious crop that was generally unhappy unless growing in a Mediterranean climate; it was one I failed with on many occasions. Despite the seductive description on many seed packets for new hybrids and traditional varieties, I stopped trying. I knew there could be no culinary love affair between us when the best choice was to purchase from a generally uninspiring selection of fruits that had lost their colour and vibrancy having spent too long on supermarket shelves.

My prejudice against the aubergine was underpinned by a belief that it was not bred to be grown in Britain. To an extent that has been true, but aubergines are grown in hugely diverse climates in their homeland of the Indian subcontinent. The plant was first domesticated in India four thousand years ago, and as of 2010 about 2,500 known varieties were under cultivation or in gene banks.[2] So finding success in my garden should surely have been simply a matter of identifying which varieties had the traits suited to growing in my corner of South Wales. Of course, they would have to be unimpeachably delicious to justify their place. As we shall see, just because there is a huge diversity of varieties, all derived from an indigenous population growing in different climates and soil conditions, there is no guarantee that there is an aubergine out there that will feel at home with me.

The fruits that Mrs Singh had in the kitchen were the size of tangerines, pale purple with faint paler stripes. Earlier in the afternoon, I had tagged along with her to the market at the end of her street to buy a few. They were redolent

195

of aubergines that I had seen growing on organic farms I had visited in other parts of Rajasthan. I had been given a packet of seed by my guide Narendra, himself a farmer. The cultivar, Vishwanath, was bred by Kudrat Seeds with the support of India's Department of Science and Technology. This is a public/private initiative helping in the development of improved varieties through participatory breeding programmes with farmers across India, strengthening resilience in horticulture and promoting organic practices as well as saving, sharing and improving seed. I was not holding my breath that the seeds I had been given would be happy back home. The aubergine exemplifies the diversity of classical and molecular breeding strategies that enable growers to create delicious new varieties that can flourish in the most unlikely of conditions. But would these innovations prove sustainable and resilient? I was to find out a few months later.

My expectations for Mrs Singh's dish were not high, but when I scooped up a mouthful of *baigan masala* in a fold of a freshly made chapati, I saw the light – or rather tasted it! Maybe aubergines were not as inedible as I had always thought.

Sometimes There Are No Seeds

Aubergine can be a challenging crop for freelance and open-source breeders because it has evolved with a particular trait, parthenocarpy, which means fruit can set without being pollinated. Aubergine pollination is sensitive to unfavourable climatic conditions, including too much or too little water and extremes of temperature, whether too hot or too cold. Although farmers need some of their crop to set seed, it is more important that they get a decent yield in adverse conditions and parthenocarpy makes this more likely. The genetics governing this trait are complex and can be seen in other crops too, especially cucumbers and occasionally tomatoes, where I find few if any seed in some fruits. Plant breeders also like to induce parthenocarpy to get seedless fruits that appeal to the consumer. It is a tool much favoured by major breeders both in creating new F1 hybrid cultivars and in some GM development, including aubergines, because farmers must buy fresh seed every year.[3] The local seeds Narendra gave me have shown a high degree of parthenocarpy and I have only been able to save a few seeds from a couple of fruit over several seasons. It has yet to love growing in my garden. Hopefully persistence, through saving seeds from the first fruit in the hope that I can select for earliness, greater hardiness and, with those traits, a great taste too, will pay off.

Although aubergine is a very important crop in the Indian subcontinent, it remains more of a niche in the UK. Commercial breeders show little interest in coming up with new varieties that might be better suited to a British climate: hence my desire to find someone who does. Fortunately, that person exists.

A Better Brinjal for Cooler Climes

It was on a wet early autumn morning, while sheltering from the drizzle in a Danish garden overlooking a slate-grey sea, that I met a man who has become obsessed with breeding a hardy and delicious aubergine that can be grown with certainty outside in his homeland – and hopefully in mine too. Søren Holt has a wonderfully curious and open mind when it comes to developing an aubergine that will flourish in his shady and often windy and wet garden. He's been at it for some time. In the autumn of 2007, he decided to

The Accidental Seed Heroes

save seeds of an American variety called Applegreen. This produces apple-sized green fruit, as the name suggests, and was bred in 1964 by one of America's great plant breeders of the twentieth century, Elwyn Meader (1910–1996) of the University of New Hampshire. It is known for its compact habit and ability to flourish in cool climates with a short growing season; as a result, it is popular in Scandinavia and was a good starting point for someone who wanted to see if he could breed a variety suitable for outdoor cultivation. The following year, Søren grew a number of different varieties, some recommended by friends and others because of where they had been bred originally: Czechia, Ukraine, Russia, Thailand and the US.

His first challenge was to get his fruits to produce seeds. Aubergine flowers are similar to those of tomatoes – they are members of the same family – with some varieties having inserted stigma, which means they are entirely self-fertile and will not cross with other varieties unless the crossing is done mechanically. Other flower types have exerted stigma, meaning that pollen can be easily transferred from one flower to another by a visiting bee. Søren wanted to create opportunities for his different varieties to cross with each other accidentally. So, in his first year he let nature take control and waited to see what, if any, crossing might take place. He would only know if he had been successful the following year, when any plants he grew bore fruit that appeared new or different.

One aubergine that especially interested Søren was Skorospely, a Russian early variety with oblong fruit. That year he had early crops from two other varieties: Rima F1, a popular hybrid commercial one bred by Syngenta, and a drought-resistant Ukrainian variety, Almaz, also known as Black Diamond, which had been collected and first sold

Beautiful Brinjal

by the Seed Savers Exchange in the US in 1993. All three varieties produce elongated dark purple fruit, some fat, some thin. Søren was especially impressed with Rima F1, but knew it would not breed true from seed. What to do? Dehybridise it. Having made no attempt at isolation, he had hoped that there might be accidental crossing between the three varieties he believed had the potential for future outdoor cultivation. He was on a journey of discovery, driven not by the systematic approach of modern breeders who know exactly who the parents of new offspring are, but by the desire to simply let nature be in charge. In doing this, Søren was following in the long tradition of farmer-led breeding. He has form. He has already been able to grow red peppers and melons outside from home-saved seed carefully selected for earliness and happiness in his garden. Not the approach of a professional, but, in my humble opinion, just as important. The result: new locally adapted Danish FVs.

I identify with Søren. When he travels he likes to return with seeds. December 2008 was no exception. He came back from a trip to Irkutsk in Siberia laden down with a great diversity of vegetable seeds, including several aubergine varieties. Unable to read Cyrillic, he was mostly in the dark about exactly what he had bought, other than hoping the pretty pictures on the packets were not works of fiction. For sure, all the seeds were in the shops because locals grew them. Summers in Irkutsk are short, hot and humid. Not exactly Scandinavian weather, but Søren felt optimistic. And among all those packets were a couple of F1 hybrids that he could try to dehybridise if his work with Rima were to prove fruitless: seed insurance, if you like.

The summer of 2009 was to be a frustrating one. Having successfully saved seeds from Rima F1 the previous year,

The Accidental Seed Heroes

Søren was able to embark on dehybridising. With the next generation – the F2s – one can expect great variability with no guarantee that any of the fruit will be tasty or, as he was hoping, a good size. Søren wanted his fruits to be big! Working with dehybridised F1s and accidental crossing to boot, it might take him ten years to select and stabilise a new variety. That summer one plant produced large, long and glossy black fruit but could the seeds saved from it be fertile? Would a next generation show similar traits?

The following September, after a cold spring and cool summer, fruit set was very poor and Søren failed to save any seed. Those from the one beautiful fruit he had had such high hopes for the previous season were sterile. But, ever the optimist, he still had seed from 2009, and he could continue to work with it the following year.

The summer of 2011 was no better than the previous year, but a number of Søren's plants produced fruit from 2009 seed that survived the wet weather. These were third-generation – F3 – Rima crosses. It is important to keep good records as you grow through the generations. Sorting the traits according to the parentage enables the breeder to identify which elements are associated with a particular mother plant. What started off as random crossing became more organised in Søren's vegetable patch. By now he had grown three breeding lines of F3 and one of them produced a number of plants with good crops of large fruit. Even if they failed to set seed, he knew who their mother was and he still had some seed left: Rima F3 No.3. Søren had been conducting his growing plan in the open ground, but decided he should grow his new generation lines in a greenhouse to have a greater chance of getting fruits containing seeds. Pragmatism over purism.

It was another cool summer in 2012, but Søren was finally able to get seed from his favoured breeding line and in large quantities. He continued to select from a number of lines until, by 2017, he had two lines that gave consistent results when grown in the open. He named them Huginn and Muninn, after the two ravens who were the eyes and ears of the great Norse god Odin. He gave seeds of these two lines to a fellow grower, Ann Marie, and after three years of selecting and saving seeds she gave Søren a few plants, as the fruits she was enjoying were a little different. In 2020 he crossed Ann Marie's plants with Huginn and Muninn and the result is a delicious, long, black fruit that shows diversity in shape – the longest fruit being the earliest. Søren gave this creation a new name, *Sorte Ravne* (Black Raven). It is the first Danish FV aubergine and I plan to grow it myself as an outdoor crop. Today, with others selecting from crops they have grown from his seeds, Søren believes there are probably three FVs growing in the milder regions of Denmark. And just how delicious is Black Raven? Well, Søren loves it and encourages others to enjoy it. For me, the jury is out until I can get a crop to harvest in a wet Welsh summer.

How GM Got In on the Act

The way GM became part of the story of aubergine plant breeding started with cotton. In 2002, India approved the cultivation of a genetically modified transgenic cotton that was resistant to a devastating pest, bollworm (*Helicoverpa zea*). Developed by Monsanto, this crop was given commercial approval in the US in 1995 and China in 1997. But Bt cotton, as it is commonly called, is so named because of the addition of genes that encode toxin crystals derived from

The Accidental Seed Heroes

strains of *Bacillus thuringiensis* (Bt), a naturally occurring soil bacterium, that is used widely as a biological control in both organic and conventional agriculture. There are hundreds of different Bt toxins, which kill the larvae of a huge number of Lepidoptera – moths and butterflies – as well as some other insects, many of which predate not only on cotton but on other crops, including maize and aubergine.

Monsanto was able to insert the active elements of the bacillus into cotton as a transgene, which meant the plant could now produce the natural insecticide itself. The company then went into partnership with one of India's largest plant breeders, Mahyco, to reset India's cotton production. By 2011 India was the largest producer of GM cotton in the world, with 96 per cent of all cotton in the country transgenic.

The results for farm economies, farmer health and biodiversity were dramatic and also highly contested.[4] Some studies on the use of GM cotton concluded that the reduction in the use of broad-spectrum insecticides showed a marked increase in the types and numbers of predators such as ladybirds, lacewings and spiders, thus reducing aphid populations. This positive impact was also shown to spill over into other crops affected by aphids, such as maize.[5]

However, other studies have shown that any gain in reduction of chemical use was short-lived and that today farmers are using even more pesticides with greatly increased populations of non-targeted pests and resistant strains of the pink bollworm that Bt cotton was meant to be resistant to.[6] Advocates of classical approaches to plant breeding and agricultural practice, including Dr Vandana Shiva – the Indian environmentalist activist known as 'the Gandhi of grain' – challenge the claims of reduced pesticide use and improved yield.[7] Monsanto's GM cotton has not, many believe, lived up to the hype. The seed is

202

Beautiful Brinjal

very expensive and farmers who grow it must use chemicals to control other types of pests and diseases, so incomes have, if anything, fallen for all but the largest monoculture farms. Plus, there is the problem of acquired resistance from the pest.

May 2019: in headline news across India, an illiterate farmer from a village in Haryana state, about 100 miles (150 kilometres) southwest of Delhi, had been growing a genetically modified aubergine. Anti-GM protestors and a lobby group opposed to the use of GM in any form in India demanded a complete moratorium on growing GM crops. The previous month, the poor farmer had been nabbed by activists who demanded he cease selling his produce. He had allegedly bought seedlings from a vendor at a bus stop in town as he had been told that the plants were resistant to a particularly devastating pest, the larva of the eggplant fruit and shoot borer moth (EFSB) *Leucinodes orbonalis*, which is native to the subtropical regions of Asia and Australia. The farmer had no idea he was buying a genetically modified aubergine, nor I imagine would he in any event have cared what its genes were. He just wanted to feed his family, grow his brinjal and not have to douse the crop with highly toxic and costly insecticides that were as likely to kill him as the bug. The alternative was to grow conventional brinjal and lose up to two thirds of the crop to the wretched caterpillar. So how had transgenic aubergines, a banned crop in India, found their way into the country?

A GM Success Story?

An indefinite moratorium on the use of Bt brinjal in India had been in place since 2010, although the government's own appraisal committee had found it to be bio-safe in 2009.

The Accidental Seed Heroes

Neighbouring Bangladesh took a different view. In 2000 Mahyco began developing Bt brinjal using the same gene that Monsanto had applied to Bt cotton. *Bacillus thuringiensis* can control EFSB in the same way as it does the cotton pests. In 2003, Cornell University, with $4.8 million of funding from USAID, facilitated a programme to help solve critical food problems, including insect damage in brinjal. The programme ended in March 2021, seven years after Bt brinjal was cleared to be grown commercially in Bangladesh and became the first GM food crop to be adopted in South Asia.[8]

The cultivation of Bt brinjal in Bangladesh was unique in that it was the result of a public and private sector collaboration on one of the most important and affordable vegetables grown in the country. Until now, GM seeds had been sold at a premium and growers had been forbidden from saving seeds, even when they were self-fertile and open-pollinated, as many types of GM soya, rice and wheat are. Monsanto and its like wanted to ensure their customers bought fresh seed every season. Mahyco saw no commercial value in developing transgenic brinjal itself and agreed to pass the licence on to the Bangladeshi authorities for free. Over the next decade the Bangladesh Agricultural Research Institute (BARI) developed breeding lines and bulked up seed: by October 2013, four lines of the genetically modified Bt brinjal – two open-pollinated and two F1 hybrid cultivars – that could be grown in different regions of Bangladesh were trialled. In the 2013/14 season just twenty farmers across the country grew Bt brinjal, some choosing to pay for hybrid seed because they believed they would get a better yield and price than from the freely available open-pollinated cultivars. Then word among farmers in the trial spread and, as more seed became available, these new cultivars that required fewer

Beautiful Brinjal

chemical interventions and resulted in greatly improved yields and less sickness became much sought after. Within three years nearly 28,000 farmers across Bangladesh were growing Bt brinjal; when the last data was released in 2021 some 65,000 of the country's 150,000 aubergine-growing farmers were planting cultivars of it and as many as 20 per cent of them were saving and sharing seed.[9] Farm incomes have improved with increased yields, revenue and, crucially, over 80 per cent of farmers surveyed said they were happy with the quality of the produce from Bt brinjal, compared with just 28 per cent of those who grew non-Bt brinjal. The amount of Bt brinjal seed being used by these small-scale farmers is increasing as word of its benefits continues to spread. But, as with all GM crops, the pests and diseases they have been designed to combat have a nasty habit of mutating and adapting themselves.

Pest resistance is a big problem, and one of the main arguments against GM as the be-all and end-all for the future of food security. Farmers have to employ strategies to slow down the development of resistance in pests and diseases, principally through refuge creation. This requires them to plant seeds of conventional non-resistant EFSB varieties in areas close by their GM crops, to encourage crossing between them. Although the resistant genes are recessive, the larvae continue to be killed when eating Bt brinjal, thus limiting the size of the resistant population. There are real challenges economically in using this method, because up to half the area under cultivation might need to be used as a refuge, seriously reducing yields and also increasing pesticide use to ensure the non-GM crop delivers a meaningful harvest. In truth, GM can never solve every problem, but once farmers have embraced it and it is out there uncontrolled, it is no longer possible to put

The Accidental Seed Heroes

that particular genie back in the bottle. And maybe the result is not all bad, as some might have one believe.

I am fascinated by the science of GM, but central to where the issues lie is how it is applied and by whom. Apart from a few notable exceptions of participatory breeding between research institutions and indigenous farmers, as we have seen in previous chapters, what drives innovations in GM is the desire among agribusinesses to dominate the seed market in support of ever-larger farms growing monocultures. Research and development of new cultivars of key crops is indeed intended to offer solutions to reduce water and chemical inputs, but it is fundamentally about ownership of seeds and I passionately believe that said ownership should be in the hands of the grower. Although Bt brinjal is freely available to all farmers without the need to buy a licence or pay a premium, the cultivars on offer represent a genetic bottleneck because they are bred from a narrow selection of germplasm held by commercial and public breeders, not farmers. They can be withheld at any time, because the farmers do not control the means of development. Despite this, for the first time farmers now have an opportunity to be custodians of any changes they effect themselves on this new GM vegetable and how all this is playing out I am taking a close interest in.

The Bangladesh example challenges many of the assumptions – made by those opposed to GM – that the very act of putting transgenic material into the world is a recipe for disaster for farmers, consumers and the planet. An argument, and one I have advocated and support, is that we don't need GM because there is enough diversity for breeders to develop new cultivars that are resistant to the pests and diseases the GM alternative claims to address. But this is not a binary argument: there are many shades of grey! India is the centre of diversity for aubergine but, despite the fact that breeders have access to the germplasm of thousands of different varieties, they have yet to come up with a new cultivar that is naturally resistant to EFSB. So maybe GM does have a part to play here.

Another of the great shibboleths among opponents to GM is that traits within the crops, such as resistance to glyphosate, can be transferred to wild populations with unknown and possibly catastrophic consequences. This is argued in India because farmers growing Bt brinjal illegally can cause accidental, or even deliberate, crossing with non-Bt brinjal varieties and thus pass on EFSB resistance in a non-systematic way. Across the border in Bangladesh, on the other hand, the government gave strong support to ensure Bt brinjal could be an economically viable crop with a robust and easily enforceable regulatory framework that put farmers first and recognised the need for an equitable public and private partnership. Rather than introducing just one new Bt brinjal cultivar, they introduced four open-pollinated and F1 versions, providing real choice to farmers who each have their own preferences and priorities.

And as with everything to do with what we eat, not just yield and resilience are important, but flavour too. Some of the GM varieties were tastier than others; some did better in different parts of the country. So what did the Bangladeshi farmers do? About 20 per cent saved their own seeds, which they also shared with other farmers. This is perfectly legal, though discouraged if practised over many seasons because of the need to avoid accidental crossing with non-Bt brinjal varieties.[10] Preserving natural enemies is also an important part of pest management in delaying and slowing down the evolution of a resistant EFSB. To date no evidence of resistance has been found. The lives of thousands of resource-poor

The Accidental Seed Heroes

farmers have been transformed and the success of Bt brinjal is evidence that the GM argument needs to move away from one between good and evil to recognise that, in this case, the debate reveals many shades of grey.

A note of caution and something that goes to the heart of the question of a sustainable future for a diverse food system involving GM: US funding to support development and deployment of Bt brinjal in Bangladesh has ended and right now there is no certainty that USAID will keep putting its hand in its pocket. There are limited, if any, financial resources available from the Bangladesh government. If Bt brinjal cultivation is to be continued in Bangladesh (and in the Philippines, which is also now licensing it as a fodder crop), superior and second-generation lines with multiple Bt genes that make them more resistant and that are adapted to different growing conditions and consumer taste need to be developed. We need to accept that transgenic breeding is not a panacea, but is based on a business model of continual tweaking – the GM treadmill. Simply banning it means demonising the curious scientific mind. GM can have a place in the diversity of approaches needed to develop crops fit for the future. Sustainable and affordable solutions that do not require external financial support should be the focus, especially when accessible tools such as marker-assisted selection could have an equally beneficial impact without the need for constant vigilance that GMOs demand.

Which brings me back to what is going on in India. Government support of the sort we see in Bangladesh is absent. The country continues to deny farmers access to Bt brinjal, which has been deemed safe by its own regulators. Those farmers look across the border and see their neighbours growing successfully, with no evidence of an environmental or ecological downside. Not only that, but Bangladeshi farmers are not poisoning themselves as they were before with uncontrolled applications of deadly insecticides, whereas Indian farmers continue to suffer. It is dangerous to presume that the benign situation in Bangladesh will continue and that pesticide use to combat other bugs and diseases won't increase, as it has with Bt cotton. A burgeoning black market now exists in India, infuriating anti-GM activists, who blame the Indian regulatory body, the Genetic Engineering Appraisal Committee, for ignoring it. The illegal trade in locally bred transgenic cotton also continues despite the efforts of licensed producers to get it stopped. But in the case of Bt brinjal, it suits the government to turn a blind eye. The economic and health benefits to farmers of growing Bt brinjal are obvious and it is one less headache for government when their farmers are making a profit!

Not all illegal seed is coming across the border. Probably very little, in fact, because what is being produced is being fully employed by Bangladeshi growers. When the 2010 moratorium was imposed in India, all transgenic material should have been deposited in a secure storage facility until further notice. It appears the seeds were not. All this finger-pointing does nothing to resolve a fundamental problem: how to ensure proper oversight of the use and development of Bt brinjal. Many fear an irreversible loss of diversity if the 2,550 varieties in India are all contaminated by cross-pollinating with the transgenic varieties – though this is somewhat unlikely in my view, especially as the germ-plasm is held within India's gene banks.

So, what are the lessons I take away from the Bt brinjal story so far and how does it fit in the world of developing aubergines that will feed us in our changing climate? Firstly, if we are to embrace aspects of GM, then it needs to develop within an environment where farmer and breeder can collaborate at a local level. It has to be affordable and strengthen the diversity of low-input and ecologically sustainable food production. The default should be participatory breeding programmes. We have seen the importance of the role of the amateur breeder with a passion to develop an aubergine that is suited specifically to how they want to grow the crop. Among all the plant scientists and breeders in India I hope there are people like Søren who, working with the region's vast germplasm, will plug away over many years to come up with an EFSB-resistant brinjal.

I like to celebrate the small, round brinjal I was given in Rajasthan, which turned my prejudice against the vegetable around, and I take my lead from Søren, who impressed upon me what matters to him – namely that his Black Raven is secure into the future. He will achieve this aim because other growers can breed their own FV from his aubergine. What is important is that breeders and amateur enthusiasts like him have access to the genetic diversity in aubergines and as he says: 'Allow Black Raven to flow a bit, simultaneously adapting to our changing environment.' My next adventure in aubergine gastronomy will be to make Søren's Black Raven feel at home in my garden.

CHAPTER ELEVEN

It's All in the Pip
Fruitful Labour for Apple Breeders

*Eat an apple on going to bed and you'll keep the
doctor from earning his bread.*
Pembrokeshire saying, circa 1866

The hedge was a model of neglect. Over 30 feet high,
ash saplings, whip thin, reached for the sky alongside
unruly hazel and the odd apple and pear tree. Something had
to be done! It was early May, and my wife and I had become
the proud owners of a handful of acres in South Wales. It
was not just the hedge that was home to the last survivors
of what, in a previous generation, had been a common sight
in the bucolic landscape: everything was in blossom and that
included two struggling apple trees, thin and feeble, a couple
of ancient specimens that were all that remained of what had
once been an orchard.

The hedge needed to be restored and that involved removing the apple trees. But work would have to wait until the
winter. This didn't mean I would lose them. I could take cuttings in January and graft them onto a rootstock to be planted
back into a new and productive orchard. In the autumn I
found a single windfall; it was delicious – a combination of
acidity, sweetness and a crisp, firm flesh – motivation if it

211

was needed to take cuttings of that year's growth. The trees were felled, the trunks milled into boards for others to turn into useful objects – I am no carpenter – and I selected a few scions I needed to give to someone far more skilled than me to graft onto a suitable rootstock.

I secretly hoped that what I was growing was from a chance seedling, an accidental cross between long-lost varieties from the original orchard. I would be the proud owner of delicious new Welsh cultivars that I could put my name to. It was to be some time before the truth would be revealed.

Apple Breeding – A Salutary Lesson?

Today's apple cultivars are the result of cross-breeding from a very small number of varieties.[1] In one sense, this is a familiar story. Modern plant breeding focuses on working with varieties that express tried and tested traits to come up with new creations that offer the breeder the shortest route to – in this case – the next great apple that can be grown at scale and sold in supermarkets everywhere.

Golden Delicious is second only to Red Delicious as the most widely cultivated apple in the world. It's Britain's favourite, is hugely popular in the US, can be endured in countless hotel breakfast buffets, bought by fans at farmers markets and, of course, picked up in supermarkets. No place seems to be immune to selling it. Spotted as a seedling in 1891 by fifteen-year-old J.M. Mullins while he was scything the orchard on his family's farm in West Virginia, he decided to let it live. The seedling was likely the result of an accidental cross between two popular cultivars in the orchard: Grimes Golden Apple, itself an accidental cross discovered nearly a century earlier, and Golden Reinette (aka English Pippin).

It's All in the Pip

The family sold the rights to propagate the apple, which they called Mullins' Yellow Seedling, to Stark Brothers Nurseries and Orchards – a major mail-order company – for the princely sum of $5,000. The Mullins' apple was renamed Golden Delicious and marketed alongside another of Stark's products, Red Delicious.[2]

There is no denying the appeal of these twentieth-century superstars of the apple world. Their colour – the red in particular – appealed to the American eye as much as their sweetness. Apples are big business and so breeders have been endlessly tweaking these varieties, often to their detriment. In the case of Red Delicious, which begat fifty variants, the later versions became so far removed from the original that customers stopped buying them.[3] Rather like the Habsburg dynasty who ruled over their Austrian Empire for centuries, inbreeding proved their, and this apple's, downfall. What mattered to those who treated apples as a commodity – just as with so many other major crops that appear in this book – were yield, storage ability and appearance: key traits to maximise profits, but at the expense of flavour. Today apples have been bred to keep for a year or more in climate-controlled storage and as a result are, to me at least, tasteless phantoms of what should be enjoyed when biting into such a wonderful fruit. This is what happened to Red Delicious. It had led the field by a mile but, by 1998, sales had fallen to the extent that it was bankrupting farmers who had chosen to grow it pretty much to the exclusion of all others. So bad was the collapse in the market that in 2000 the US government bailed out the industry, which had lost some $700 million in the previous three years.[4] Discerning customers had made their feelings clear, but in so doing had they stepped from the fireplace into the fire?

The Accidental Seed Heroes

Red Delicious was toppled from top spot by Gala, the result of a deliberate breeding programme in New Zealand in the 1930s of a cross between a locally bred apple, Kidd's Orange Red, and Golden Delicious. Having been granted a US patent in 1974, Gala arrived in the UK about forty years ago and is now traded under the brand name Royal Gala. It remains one of the most popular apples sold around the world. Traditional and delicious varieties that were bred to flourish in the UK, such as Egremont Russet or Worcester Pearmain, are now a rare sight on the supermarket shelf, replaced by popular cultivars from the southern hemisphere. One of the most significant of these is Braeburn. Discovered by a farmer outside Motueka on New Zealand's South Island in 1952, it was cultivated for export. Braeburn's parents are Golden Delicious and an English apple first identified in 1827, Sturmer Pippin.* This identification was made possible thanks to plant scientists mapping the genome of Golden Delicious in 2010.[5]

A Battle for Domination

So, who or what are the 'new breed' of apples that dominate sales? One is Jazz, a cross between Gala and Braeburn, developed in the 1980s. Another is Pink Lady, a cross between Golden Delicious and another Australian cultivar, Lady Williams, itself a seedling from an accidental cross between another famous Australian apple, Granny Smith (much loved in the UK), and an American cultivar from the 1830s, Jonathan.

---

* Sturmer Pippin was exported to the UK from Tasmania and New Zealand from late in the nineteenth century.

It's All in the Pip

These apples aren't cultivars; they are brands. Jazz is the trademark brand of a cultivar called Scifresh and can only be grown under licence. The same goes for Pink Lady, which is the trademark of a cultivar called Cripp's Pink. Pear and Apple Australia own all rights in Pink Lady and have forbidden it from being grown in the UK because they fear the climate will not ensure the organoleptic qualities apple-eaters apparently now want more than ever: sweetness, crunchiness and smooth skins. The consequences for British apple growers are dire when the most popular apples are all imported 'brands'.

The fact is, almost all apple cultivars are controlled by marketing and distribution companies like Worldwide Fruit who ensure their brands dominate the apple aisle in your local supermarket. In the UK the majority of apple orchards are owned and operated by British Apples and Pears Limited (BAPL), which grows about twenty-five varieties, of which just nine can be considered truly British in origin.[6]

It can take decades to breed a new cultivar of apple and as a result innovation is reliant on public funding: production is in the hands of a small number of giant growers who through licensing have complete control of all aspects of the fruit as a brand. But classical approaches to breeding and serendipitous discoveries are offering real solutions to escaping the genetic bottleneck of modern breeding approaches.

A Fabulous Mixed Heritage

The apple (*Malus domestica*) is the most important commercial fruit crop in the temperate regions of the world and the fourth most widely grown after bananas, oranges and grapes.[7] Apple trees are a common sight in the traditional British garden and are the most popular fruit tree with gardeners. The apple's domestication and cultivation have a long and distinguished history.

Its wild parent *Malus sieversii* is endemic to the Tian Shan mountains of Central Asia. Domestication happened over millennia, between four thousand and ten thousand years ago. The huge diversity of apples we have today is thanks to our ancestors, who took seeds with them as they travelled across the northern hemisphere. During this period significant changes in the species occurred because of introgression – the transfer of genetic material from one species to another – which resulted in apple populations becoming more like the European crab apple *M. sylvestris* than their other wild parent, *M. sieversii*. Hybridisation and introgression also took place between the Siberian wild apple *M. baccata* and another species from the Caucasus, *M. orientalis*.

I love a crisp, slightly sharp, firm eating apple, but over two thousand years ago the Chinese were selecting for a softer type. This gave the world two more species, *M. asiatica* and *M. prunifolia*. Thanks to modern genome sequencing, they are believed to be the result of hybridisation that took place between *M. baccata* and *M. sieversii* in Kazakhstan at that time.[8] Some wild apples like *M. sieversii* are similar in size to many modern cultivars. This made life easier for those first growers, who were selecting for traits that included degree of acidity, colour, firmness/softness and sweetness. Size mattered too, because a larger fruit looked nicer and you needed fewer of them for nutrition.

Today, all the apples we eat and the trees we buy to grow in our gardens are clones, created by grafting scions of the desired variety onto a rootstock that determines the vigour or ultimate size of the tree. Rootstocks are as important as

It's All in the Pip

the varieties that are grafted on to them because, in addition to vigour, they impart other traits such as increased precocity (early fruiting), disease resistance and productivity. There is some disagreement as to whether the Romans invented grafting as a way to scale up apple production, because a precise starting point for this technique has yet to be determined. However, propagating from cuttings was widely practised across the Mediterranean over two thousand years ago and modern genetic sequencing has enabled researchers to ascertain that some of the earliest known French varieties would appear to have Roman ancestors.[9]

It's a Numbers Game

Since the Romans named six varieties, breeding around the world has resulted in apples being among the most diverse of all edible crops. There are over 7,000 named varieties under cultivation worldwide: over the centuries, 2,500 of them have been developed in the UK. Some, but not all, of this diversity is maintained in collections and inventories of commercial fruit tree growers. In the US, the Fruit, Berry and Nut Inventory listed 1,469 apple varieties at the turn of this century. The orchards of the Plant Genetic Resources Unit (PGRU) of the USDA in Geneva, New York, are home to over 6,000 unique accessions representing 55 species and cultivated hybrids, including nearly 3,000 varieties that are used for evaluations and research. In the UK the National Fruit Collection, a collaboration between the University of Reading and DEFRA, has over 2,000 apple varieties, maintained at Brogdale Farm in Kent. National collections all around the world hold thousands of varieties that have been bred locally. Many are available as slips or young trees to be propagated and grown in our gardens and orchards.*

Despite all this diversity, commercial cultivation uses just a tiny percentage of available cultivars: of the 30,000 accessions that are widely accessible around the world just 30 are grown for sale internationally.[10] The source breeding material is literally just a handful of cultivars – known as founding clones – whose dominance varies from one region to another. These same named cultivars occur again and again in the breeding history of new ones: Golden Delicious, Cox's Orange Pippin, Jonathan, Red Delicious, McIntosh (especially in North America). A few others also make frequent appearances: James Grieve, Rome Beauty and Wealthy. In research conducted in Canada towards the end of the last century it was determined that 73 per cent of all new cultivars were related to at least one of five founding clones.[11]

This concentration of plant inbreeding has serious downsides. It has been known for the last 70 years that inbreeding increases uniformity – very important for supermarkets – but at the expense of other traits, especially flavour. Perhaps more importantly, using such a small gene pool means that pathogens – especially new ones for which there is no genetic resistance within the founder clones – require costly

---

* Descriptions of apples as varieties or cultivars can be confusing because the words are frequently interchangeable. For the avoidance of doubt, a variety is one grown from a seedling, be it wild or an accidental cross. It becomes a cultivar when it is propagated and marketed under a particular name or when it is the result of deliberate crossing. A cultivar should not be confused with a brand, which is a marketing name only. Pippin in the name means it came originally from a pip, as with Cox's Orange Pippin.

It's All in the Pip

spraying regimes. As a result, in the last decade residue from the application of pesticides and herbicides that we find on fruit we buy has more than doubled.[12] Evidence from fifty years ago confirmed reduced vigour and survival rates of new cultivars bred from Cox's Orange Pippin.[13] There is a chronic lack of research into the long-term effects of inbreeding although, as I reveal later, breeding approaches to create the next generation of mass-appeal apples recognise the need to introduce genes from heritage and wild apples to reduce pesticide use and improve flavour.

All the vegetables I have written about in this book are self-compatible – they inherit their genes from both parents and can self-pollinate. Apples are different: they are self-incompatible and must cross-pollinate to develop fruit. They are also an example of a plant with 'extreme heterozygotes'. This means that, rather than inheriting genes from the parents, which results in a new variety with similar characteristics, the offspring is significantly different to both. This is why orchards need to contain apples of the same group or ones with a similar flowering time. Heterozygotes means the plant breeder has no idea what they are going to get until they harvest fruit from plants they have grown from their saved seed.

Our British winters are warming up. Fifty years ago, we had twenty-one more days of frost in a year than we do today.[14] In order to trigger the development of flower buds, apple trees need to have a period of dormancy induced by frosty days and nights – measured as chill hours, when the mean temperature is below 7°C between 1 October and 1 April. Some modern cultivars, such as Granny Smith, only need two hundred chill hours (although they need long warm summers too), whereas the most popular cultivars need between five hundred and one thousand hours – five to six weeks in midwinter. As our

219

winters are getting shorter and milder, the reduction in chill hours is affecting apple trees from flowering and setting fruit effectively. Researchers at the University of Reading have noted a considerable reduction in chill hours in just the last few years – down by three hundred hours.[15] Traditional British apples, especially, could become a thing of the past because they were bred when there was no shortage of chill hours and only modern brands from New Zealand and Japan with short chill hour requirements will yield meaningful and reliable harvests. My interest in growing local varieties that were bred one hundred or more years ago could be a mistake. They might do better in the north of the country or Scotland and I need to find cultivars that do best two hundred miles due south in northern France. Efforts to find new accidental crosses from wild and older traditional varieties that are more resilient and able to flower well after a mild winter may, as we shall see, start bearing fruit!

Crowd Breeding to Create Apple Forests

The University of Copenhagen is home to a gene bank of about eight hundred apple cultivars, many of them unique to Denmark and other parts of Scandinavia. In 2013, two scientists working in the Department of Plant and Environmental Sciences, Maren Korsgaard and Torben Bo Toldam Andersen, asked themselves the questions, 'Why should Kazakhstan be the only country to have apple forests?' And, 'Why can't Denmark have them too?' They started a programme – Apple Oasis – to spread apple genes far and wide across the country in order to see what Mother Nature might come up with. Importantly, they wanted to develop new and robust cultivars that would be truly local.

It's All in the Pip

Apple breeding is a costly and time-consuming endeavour, so how to do it without spending loads of money? Maren and Torben selected seeds from mother trees: fifty-four of the most robust and delicious old Danish varieties that showed resistance, when grown organically, to three serious diseases – apple scab, mildew and canker. Volunteers were engaged in an innovative and long-term project – a participation between growers and plant scientists; a marriage of traditional cultivation and DNA analysis to identify both parents.

In the Danish experiment, ten thousand seeds of the fifty-four cultivars were distributed among 136 volunteer gardeners and farmers. Although Maren and Torben knew who the mothers were, they had no idea of the fathers, which could have been any one of the eight hundred varieties being grown in the collection at the university. These included known cultivars from other parts of Europe and the US, wild apples from Kazakhstan, ornamentals, crab apples and other species from around the world, including old English cider apples.

After six years, a third of the 136 original volunteers had succeeded in growing a total of 1,700 new apple trees. Thanks to late spring frosts, only 71 of the 121 trees that flowered that year – 2019 – produced fruit; 30 were like crab apples and 41 produced normal-sized fruit. Out of the 1,700 trees, just over half showed little or no infection with scab, canker or mildew, but about a tenth suffered from fungal disease. A total of 33 trees were selected for evaluation, of which 13 had a good or very good taste, 7 had an average taste and 13 were considered pretty inedible. Scions from 91 of the most interesting seedlings have been grafted onto commercial rootstock and are now growing in the university's orchards for further study.[16] Bodil, one of the volunteers I had met previously, had no apples ready for me to try when I visited her in the autumn of 2023, but assured me that she had a couple of 'good trees' in her lovely orchard.

Participatory breeding is to be celebrated and needs to be at the heart of a future for sustainable innovations. Fortunately, networks of institutions across Europe have established InnOBreed* to do just that. At the time of writing, one of the Danish seedlings is being put into a trial with a Spanish partner, and not because of the deliciousness of its apples. It will be evaluated as a potential rootstock because it is unloved by rodents and, for someone like myself who has to do battle against rabbits nibbling the bark of young trees, rodent-resistant rootstock sounds like a dream come true.

A Curious Englishman

When Tom Hartley has escaped his day job in Bristol as a certification officer for the Soil Association, he likes to map the locations of self-seeded wild apples in and around the city. He has noticed countless wild ones growing along the nearby motorways and presumes they are probably to be found across the country – all thanks to drivers and their passengers chucking apple cores out of the window! The fruits of some of these self-seeded apples have excellent flavour, he assures me. Many have been growing for years and appear healthy, with good crops of autumn fruit. Perhaps the next great British cultivar is waiting to be discovered off the hard shoulder.

---

* The work of InnOBreed to support research and development into organic crop breeding can be found here: https://innobreed.eu.

Tom's family has a large garden in Surrey where he is growing about 160 apple trees. They are a mix of seed he collected from the forests around Almaty, Kazakhstan, in 2012 and seed he received from the apple collection held by the USDA Plant Genetic Resources Unit (PGRU) – which, by the way, is home to one of the world's largest collections of *Malus sieversii* from Kazakhstan. Now Tom's trees are growing, with varying degrees of success: to date, he tells me, three or four are very promising eaters. Next phase is to graft the most interesting cultivars onto different rootstocks. He might also graft scions from some of the wild apples discovered on his foraging trips around Bristol. In the coming years he will know whether his passion for growing seedlings has resulted in a fabulous new cultivar we can all enjoy eating.

An American Answer

In the last two hundred years, the Americans have been as active and influential in the commoditisation of apple cultivars as the British and New Zealanders. On my many happy autumnal road trips along the byways of the East Coast states, I would frequently stop at a roadside stall selling 'apple cider'. The beverage came without alcohol and was generally somewhat sweeter than my favourite apple juice, which I buy from a farmer in my home county, Monmouthshire.

Thankfully, for my palate at least, there has emerged a great desire among many small-scale US producers to make what I call 'proper' cider – fermented apple juice. Until recently, the few brewers in the US making cider followed in the European tradition, using only European cider-apple cultivars. But since the mid-2010s, growers and cider makers have wanted to develop a pomona* of new North American varieties, almost exclusively selected from wild seedlings. In 2019 a young sheep farmer and orchardist from Williamsburg, Massachusetts, Matt Kaminsky, started the Annual Wild and Seedling Pomological Exhibition. The event showcases the fruit of wild seedling apples and pears that have been foraged from across the US. In 2023, over 120 fruit, which came in a kaleidoscope of colours and shapes, all delicious in their own unique ways, were tasted, commented on and, in the case of cider, drunk.[17] Kaminsky, like so many new-generation breeders and growers, is driven by a passion to make beautiful cider using American apples. He is just one of a diverse cohort of freelance breeders, apple growers and institutions that are on a mission to restore and celebrate apple diversity, both as an alcoholic beverage and for eating.

One such institution is the University of Idaho's College of Agriculture and Life Sciences. Since 2019 they have hosted the Heritage Orchard Conference, a series of webinars that are seen by thousands of interested people, myself included, from twenty-seven countries. A mix of practical growing and breeding sessions, much of the discussion centres on how to breed apples, with a focus on three key customer requirements for sweet, eating apples: deliciousness, consistent fruit quality at the time of harvesting, and storability. Breeding targets, for both sweet and cider apples, are: improved disease resistance, good appearance, the ability to flourish in a changing climate in low-input agroecological systems and breeding out both immature fruit drop and the habit of apples to crop every other year!

---

* Pomona was a wood nymph, the Roman goddess of fruitful abundance; in the context of apples, the term refers to a collection of different cultivars or varieties from a specific region.

It's All in the Pip

Sticking With What You Know

Despite so much going on in the world of apple breeding, there are wonderful stalwarts I would never want to 'improve'. I only have a very few apple cultivars in my garden – all traditional. When my wife and I moved to Wales a magnificent and elderly specimen, Annie Elizabeth, stood sentinel in what had been an orchard. Raised from seed by Samuel Greatorex from Leicestershire in 1857 and named after his daughter, it received an RHS First Class Certificate in 1868 and was sold commercially from the late 1890s. We loved this reliable and heavy cropper, with its sweet and slightly acidic flavour and beautiful white flesh, which was as delicious eaten fresh from the tree as when cooked to make apple crumble. Blackbirds and winter-visiting fieldfares loved it too and would gorge themselves on the fallen fruit. I could pick from the tree into December and store fruit well into the New Year. Sadly, a winter storm in 2019 saw its demise. Thankfully, many nurseries in the UK continue to propagate this classic from the Victorian age, so Annie Elizabeth shall return. However, needing many chill hours to bear fruit, she probably needs to be grown two hundred miles north of me to be at her best.

Early ripening is an important trait that many breeders seek. For cider makers it can extend the harvesting season and for people like me it means I can eat a freshly picked apple in August. A prime example of this is Discovery. It was raised in 1949 by a Mr Drummer, who worked on a farm in Essex, then propagated by J. Matthews, a local nurseryman who called it Thurston August because of the month in which it could be eaten. It was renamed Discovery in 1962 and has become the go-to first eating apple of the season – and for good reason. It has a lovely firm and crisp texture with a

The Accidental Seed Heroes

wonderful balance of sweetness and acidity. But it doesn't keep, so has to be eaten as soon as possible after picking to get the best of its yumminess. And therein lies the rub: how to breed an apple that ripens as early as Discovery but can keep as well as Annie Elizabeth.

In 2021, Dr Kate Evans at the University of Minnesota defined the three key elements in commercial apple breeding as: to use the natural diversity within the species with controlled hybridization and targeted selection. As Sarah Kostick, an apple breeder who studied under Kate Evans, says in a webinar, she is trying to leverage the natural diversity in the species to create a package an apple-eater desires while at the same time removing the undesirable aspects of the diversity.[18]

The most famous apple bred at the university is probably Honeycrisp, which remains hugely popular in the US. It begat a number of cultivars that are trademarked under the brand names Sweet Tango, First Kiss and Rave. Another is Triumph, bred for resistance to apple scab, which should make it of interest to organic growers who don't want to drench their crop in fungicides. Targeted, commercial apple breeding is not for the impatient. In selecting which varieties to cross, Sarah Kostick uses both observational and molecular data. It can take up to thirty years to bring a major new cultivar to market, with thousands of seedlings discarded in a continuous process of trialling and evaluating to arrive at a single apple that can compete against the incumbents. Agribusiness cannot operate on that timescale. This is why involving public institutions can benefit the giants in the apple business as well as the independent, small organic grower. The Danish approach of uncontrolled hybridisation illustrates the key difference between the means by which

It's All in the Pip

we achieve, on the one hand, a branded and trademarked commercial cultivar like Royal Gala and, on the other, an accidental cross that just happens to taste great, ripen early and store well, and was discovered in the corner of an orchard by an eagle-eyed enthusiast.

The future for apple breeding exists on two fronts. Overwhelmingly it is based on a model that starts with a relationship between institutions like the University of Minnesota and the commercial nurseries to whom they license their successes. Their customers are farmers – many of whom cultivate giant apple monocultures, as we saw with Red Delicious. The systematic approach, adopted not just in the US but in many countries where apples are an important economic asset, ensures our supermarkets are full of the same types of apples every day of the year. As customers, we rely on a model of cultivation that depends on high inputs of chemicals and fertilisers, yet does nothing to improve the diversity of apples that can bring greater natural disease resistance, and an ability to flourish in different soils and climates. On the other hand, amateur and freelance breeders are now the custodians of evolving diversity. The reality is that all breeders are in pursuit of the same three goals: deliciousness, quality at the point of sale and keeping ability. It may appear perverse and contradictory, but I am on both sides.

Never Let the Truth Get in the Way of a Good Story

Wales has a long tradition of apple cultivation, with cider being a key motivation for preserving a diversity of local varieties. Today in my neck of the woods, nurserymen like Ian Sturrock are ensuring heritage Welsh apples are maintained for us all to enjoy. He claims that these local varieties are well adapted to the Welsh climate. I think he follows in the footsteps of many a nurseryman before him in making claims that appeal to national pride, because English varieties can also be remarkably resilient in a climate that is not so different to many other parts of these hallowed isles. The culture of preserving and celebrating local varieties, so well explored by Carwyn Graves in his book *Apples of Wales*, is being restored by a number of passionate growers and breeders who are not only working to maintain what already exists but also breeding new varieties of distinctly Welsh apples, both from wild seedlings and deliberate crossing of the thirty-five or so known historic cultivars.[19]

The real challenge is to disentangle genuine cultivars from the many seedlings of variable quality that have been given names and stories that appeal to those of us who like a good yarn. For example, a 'rediscovered' native apple called Welsh Pomona came to light when hundreds of Welsh apples were DNA barcoded to determine their provenance and parents. Many had unique DNA sequences but, to date, it has been impossible to determine their parentage because so many would have been from wild seedlings or sports of old varieties that travelled freely over the border from England and were simply given different names. Some may be genuine lost cultivars. The Welsh Marches Network, an organisation of apple enthusiasts, has listed two hundred cultivars associated with the region. Apples have a long history of multiple names; nurseries were no different from other plant breeders in taking someone else's work, renaming it and claiming it for themselves. There are some in Wales today selling cultivars with Welsh names that are anything but Welsh. A good example is Cox Cymraeg: DNA fingerprinting proves that it

It's All in the Pip

is a French cultivar, Belle de Boskoop. But far be it from me to spoil a good story about the 'discovery' of Cox Cymraeg: after all, sellers of apple trees have a right to a living and now this cuckoo of a variety is part of Welsh apple history.

A Flavoursome Future for Cider

Why do apples belong in this book? I don't breed them, nor do I hunt them down on roadside verges or in abandoned orchards, but I do like eating them and drinking cider brandy – Calvados – or better still Jack High Cider Spirit, made in a 350-year-old distilling house in Dymock, a few miles from my home. Apple breeding illustrates the huge potential for a diversity of breeding strategies that can ensure we enjoy this wonderful fruit in all its manifestations as our climate changes.

Today, research in commercial breeding is almost entirely about eating apples; modern cider-apple breeding is in its infancy. But it wasn't always so. Academic research into cider making began in 1893 when Robert Neville Grenville, a farmer living near Glastonbury, received financial support from a local society to conduct experiments. So much interest did his work elicit that a decade later growers and producers across the southwest of England, including my own county of Monmouthshire, received government funds to establish a plant breeding and research centre: the Long Ashton Research Station (LARS), just outside Bristol. Its mission was to support the cider industry across the region.[20] It subsequently became the Department of Agricultural and Horticultural Research at Bristol University. What followed is a sad tale of government ceasing to support research that helped maintain diversity and resilience in British agriculture.

The Accidental Seed Heroes

In 1931, it formed the Agricultural Research Council (ARC), that took over managing LARS. Fifty years later, in 1981, the major apple-breeding section was disbanded, marking the end of innovations in the development of new cider apple cultivars. In 2003, after a century of critical work, LARS was closed for ever.

Knowing what we already have is as important as working on breeding new cultivars because flavour is of a different magnitude in heritage apples. Among the thousands of cultivars, sports and wild specimens maintained in collections around the world are individuals that express resistance to different diseases and show the ability to grow in different climates with a huge range of colours, shapes and sizes. Studying their genotype through DNA fingerprinting means we can identify rare varieties with genes a breeder might be interested in – and a cider maker in maintaining. Across the UK there are still many traditional cider orchards and collections; one in particular is in the Royal Horticultural Society's garden at Rosemore, Devon. DNA from their collection, and from lots of other orchards in the region, is being gathered as part of a major research project at Bristol University's School of Biological Sciences.

Devon is famous for its cider and this work is a collaboration between a local cider-maker, Sandford Orchards, which sources apples from a number of nearby farms as well as its own, and Bristol University, which has picked up the mantle left behind with the end of LARS. Over several years the research will establish the importance of LARS' work on new cider apple cultivars and rediscover the parentage of the ones it developed that became known as 'The Girls', whose origins were lost when LARS closed down.[21] For cider-maker Barny Butterfield, learning from the university's

230

It's All in the Pip

analysis of his own apples that his orchards contain a greater diversity of cultivars than he realised, of which many are unique, is helping him identify which trees to propagate to enable him to create more delicious cider using well-adapted and locally bred apples.

DNA fingerprinting can pinpoint so-called 'survivor apples' – cultivars that are growing in more than one orchard. Their presence indicates that they have been grafted and shared – or sold – because they have qualities that are particularly highly valued: flavour, disease resistance, storability, early ripening – the same traits breeders like Sarah Kostick at the University of Minnesota are looking for. It's exciting work, because the solutions to these breeding challenges are all to be found in existing germplasm and the new additions that are the result of the work in Bristol.

A Sweet Disappointment

My scions were grafted in midwinter, and I had to wait until the following autumn before I could plant up the maiden trees of the specimens I had discovered in 2014. It would be another five years before I got to taste the first fruit from those old specimens. I picked a few leaves from the first growth and sent them to the National Institute of Agricultural Botany's (NIAB) East Malling Research centre in Kent to be DNA fingerprinted for identification. I fantasised that the plants I had 'rescued' from an overgrown boundary hedge would be unique unnamed varieties, native to my corner of South Wales. Well, I was wrong on that count. Unique, no. But local? Yes. One apple turns out to be Brith Mawr, a handsome and famous nineteenth-century variety from Newport, the Welsh city just down the road. The other is a

The Accidental Seed Heroes

sport of Bramley's Seedling – a very common but delicious cooking apple.

What matters to me about the future for apple breeding is that more single-minded enthusiasts, alongside a new generation of traditional plant breeders, are able to increase the diversity of varieties that have a cultural value as well as an economic one with a focus as always on deliciousness. I have no desire to eat a trademarked apple with its polluting little sticker reminding me of its name alongside the ubiquitous ®. But, as I hope this chapter has shown, apple breeding is advancing on many fronts and long may it continue. And you, dear reader, can join in. Why not experiment with growing apples from seeds, maybe by collecting fallen fruit from old orchards and hedgerows? If you only have a balcony you can sow your seed in a large pot and see what emerges. Patience is needed because it will be at least five years and probably longer before the first blossom: the exercise is redolent of buying a lottery ticket. More than likely the resultant fruit will be pretty inedible. You might be able to make a passable cider, but probably nothing more. On the other hand, if your numbers come up, you could have the next Red Delicious!

Conclusion
Holding Truth to Power

As land is improved by sowing it with various crops in rotation so is the mind by exercising it with different studies.

Pliny the Younger, *Letters* 7.9

I magine it's 2050 and all the crops we grow contribute to improved biodiversity, food security and human health; they also sequester more carbon than they produce. There are plenty of reasons to believe this is nothing more than the pipe dream of a deluded old fool, but in meeting with those among us who are focused on making the pipe dream a reality, I am not of a mind to slit my throat just yet! The places I have visited, the farmers, plant breeders and activists I have met while researching this book represent a tiny minority of amazing, inspiring and important stories from around the world that give me a true sense of optimism and hope that, on my ninety-ninth birthday – assuming I am still alive to enjoy it – a paradigm shift in how we approach plant breeding in all its diversity will have taken place, and for the better.

But we have a hell of a fight on our hands. Change can only happen when we win the argument and bring the sceptics and doubters on board as collaborators. We need to listen a lot more to the range of voices: from indigenous farmers chaperoning and celebrating their evolving crops to the plant scientists whose creative minds are working at the cutting edge of genomics. To get policy change at national and international level, breeders and growers have to demonstrate success in alternative approaches. This applies in particular with cereals, where a robust alternative to the laws on DUS must be seen to work for the common good. More funding is needed to support innovation in classic plant breeding at all levels because research is becoming more competitive, and the great challenge is in scaling up proven alternative breeding and cultivation models. There has been a degrading of research into independent breeding, although this is changing as the penny begins to drop among academics and breeding institutions that the molecular approach is not a panacea.

Reasons to Be Cheerful

The world needs more farmers and institutions collaborating in participatory programmes to develop all the crops we love to eat: programmes that also include the mavericks, obsessives and people with a passion to add to the diversity and resilience with new and improved cultivars. It is freelance breeders, amateur gardeners and plant scientists, with their side interests in breeding, who I believe are making the greatest contribution. There really can be a fabulous future for diversity in plant breeding and with it a less polluted and healthier planet. And we can all join in. All of us can be part of the solution through the choices we make as to what we eat and, if we have gardens, what we grow. If we fancy adding to the diversity of edible crops by saving and sharing seeds, we can all be plant breeders too. The ways people are able to connect with their and others' food cultures and at the same time strengthen agricultural communities through

Conclusion

a recognition of the place of traditional plant breeding can be transformative.

Is this just fanciful thinking? I don't think so. I am witnessing the change when I meet farmers who are passionate about their varieties and see their future as being bound up in sharing their knowledge and produce with visitors from home and abroad. And this change is happening where it is needed most – in places threatened by but not yet fully in hock or wedded to a twentieth-century model of homogeneity: the Global South.

A grass-roots change is happening in Andhra Pradesh, known as the rice bowl of India. The regional government has recognised that the agricultural system has to change because conventional methods are poisoning the land, and the people and farm incomes are ever more insecure because of unreliable monsoons and a dearth of public sector investment. The ambition is for the entire state to be growing food agroecologically and work is well under way with the creation of Agroecological Living Landscapes (ALLs) in two districts with a history of traditional farming systems. Promoted to farmers as natural farming practice, in 2018 the government planned to transition six million farmers and their land to an entirely chemical-free system of production by 2024; at the time of writing, only around a million farmers were enrolled in the programme. However, according to Swati Renduchintala, programme manager for Andhra Pradesh Community Managed Natural Farming, who spoke to me about the creation of ALLs at the Oxford Real Farming Conference in 2024, they have seen an 11 per cent increase in productivity and a near 50 per cent increase in income as a result of embracing sustainable, agroecological farming methods that have also benefited

The Accidental Seed Heroes

their communities, social cohesion and wellbeing. Andhra Pradesh has shown us that agroecology is just as productive as intensive, high-input farming, but provides greater resilience and stability for farmers and the communities in which they live. It has also shown much-increased environmental wins and avoided negative health outcomes with improvements to income and mental health, higher levels of nutrient-dense food and the restoration of biodiversity as well as crop diversity. But success requires patience and the avoidance of unrealistic timelines.

We can use our greater understanding of the genome of our crops to improve diversity, resilience and adaptiveness. When I met the geneticist Dr Simon Griffiths at the John Innes Centre (JIC) in Norwich, he had just completed leading on sequencing the genomes of arguably one of the most important collections of landrace or farmers' variety (FV) bread wheat in the world. Assembled by the botanist and pioneering plant geneticist Arthur Ernest Watkins a century ago, some of the varieties have been lost over time. But the JIC, thanks to public funding, has been able to maintain 827 lines as a living collection. These are being embraced by researchers and plant breeders around the world to develop a new generation of diverse and locally adapted cultivars that will provide good yields in ever more challenging growing environments.[1]

Some lines in the Watkins collection came from the VIR in St Petersburg, but most were from what was then the British Empire. Simon told me that modern bread wheat is a mosaic of two groups of ancient FVs and that there are five genetically isolated ancestors that have not – as yet – contributed to modern cultivars. Many of the varieties in the collection have traits that have been lost or bred out of

Conclusion

modern wheats; the result of a drive to develop homogeneous, high-yielding cultivars. These can now be added back into breeding programmes. Simon says the collection provides answers to a sustainable future for wheat production using genomics and molecular methodologies: a toolkit that can chime with classic approaches involving evolutionary and population breeding, which I discussed in chapter 3.

At the heart of all approaches to plant breeding today there needs to be a focus on climate adaptation and a diversified agroecological landscape. Plant breeding must fit into the positive impacts of agroecology. Regenerative agriculture needs to grow enough food to feed the planet and we know it can be done. Andhra Pradesh illustrates how local adaptation benefits farmers without forcing them into endless servitude, and the work of the JIC demonstrates that conventional and molecular breeding can work together.

Once again, seeds need to be recognised as a public good and I hope that by 2050 the laws on patenting and IP will have been consigned to the dustbin of history. The freelance and independent breeder has to be able to create their new cultivars on a level playing field and to be rewarded with equitable plant breeder rights that make their work financially viable. There is a myth, propagated by multinational seed companies, misguided institutions and, worst of all, governments, philanthropists and NGOs that should know better, that the solutions to food insecurity are all technological. Molecular breeding is not the complete answer, but can be a significant part of it when not employed to perpetuate food as a commodity. All forms of breeding have their place, and they need to function in an economic climate that democratises seeds; not one that concentrates all the power in the clutches of a handful of multinational corporations

The Accidental Seed Heroes

and their acolytes. The voices opposing a narrative of inevitable transition away from the farmer to the corporation are getting louder. Farmers need to be put at the centre of innovation and we will see this evolution in the coming years – it is already happening in many places around the world – with co-innovation, participatory breeding programmes and a greater emphasis on training the next generation of plant scientists in participatory research.

The fight to limit the power of patents is going to be the most difficult to win. The European Parliament is in the process of approving legislation that will further deregulate the application of new genomic techniques (NGTs) – which give us GMOs – and their patentability. But there is pushback from some member states who see the threat to independent breeders this poses, so one can but hope the legislation will be modified or, ideally, abandoned. What is urgently needed now is unrestricted access to germplasm and NGTs to create a more equitable system of managing IP. Leading this opposition in Europe is No Patents on Seeds, a coalition of organisations involved in the conservation of the genetic diversity of edible crops.[2]

One aspect of the abuse of patent law is the actions by big seed companies to deregulate NGTs by offering free licences. This means that the farmer may not have to pay a fee to use the patented seed, but remains dependent on the breeder, who can, at any time, demand payment or withdraw the cultivar. The model perpetuates dependency, with all breeding decisions led and made by multinationals, further commodifying and controlling supply and reducing diversity and choice. What, at first sight, might appear altruistic is anything but. Licence-free seed doesn't liberate farmers from the hegemony of a vertically integrated system of production that starts with

Conclusion

patented seeds and all the inputs needed to ensure a profitable harvest. Countless patents for NGTs have been filed, not just with the EU, but in the US and around the world. Even if and when lawmakers make NGTs unpatentable, vast numbers of major crops like maize, wheat and many staple vegetables including lettuce will remain protected by existing patents.

Decriminalising the production and dissemination of seeds of heterogeneous varieties encourages further development of diverse and resilient crops and offers alternative and desirable income streams for all stakeholders. The primary lesson from the past is that an obsessive and myopic focus on a solution based solely on the latest innovations in genetics, coupled with the madness of a world that entitles breeders to claim ownership of genetic material at the expense of innovation and diversity, is not the solution. It fails to embrace the many other approaches to breeding and crop improvement that are demonstrating effective and real solutions to building a sustainable future for growing our food. The marketing and PR hype of the major breeders and their power lobbying governments, alongside bodies like the World Health Organization (WHO) and philanthropic institutions like the Bill and Melinda Gates Foundation, has to be constantly challenged. We should stop focusing on solutions through the hubris of a Silicon Valley technology-driven and monopolistic mindset: developing and evolving crops fit for purpose over the coming decades must respect people, planet and place.*

---

* Towards the end of 2024, African faith leaders wrote an open letter to the Gates Foundation demanding reparations for causing extensive damage to Africa's food systems with its support for industrialised agriculture, much of it in support of the Alliance for a Green Revolution in Africa (AGRA).

The Accidental Seed Heroes

I am frequently asked my opinion about GM, usually by people who fear the negative impact of GMOs on the environment. My answer is simple: molecular science is amazing and I welcome those curious, brilliant and enquiring geneticists who are using all the tools at their disposal to discover more about the wonders of genetics and how this knowledge can be applied to improve food security. But at present the fruits of their labour, the GMOs that now dominate production of the four most important crops in the world – maize, rice, wheat and cotton – are doing nothing to achieve this because homogeneity is degrading our soil and environment. What classical and molecular breeding have in common is the modification of a plant's genome and phenotype. But what differentiates NGTs from all the other ways to develop and improve crops that I have written about is that they exist to support a system of food production that depends upon giant monocultures, consolidation of farmland and its industrialisation, and control of plant breeding and the inputs supplied by agribusiness. Yes, some GMOs, like Bt brinjal, can be useful in reducing insecticide use in the short term, but the application of pesticides and herbicides is heaviest on large farms growing monocultures. If we want to make it easier for farms to get ever bigger and more mechanised, to continue to follow a model promoted by governments in awe of agribusiness, then great, but it is not a vision that turns me on! All monocultures create the perfect conditions for pest evolution. Remove one type of pest and for sure a different species will move in or the targeted one will mutate, making any genomic resistance ineffective.[3]

I don't envisage a world of only small, diverse farms. The great grain producers of the world – Ukraine, Russia, Canada, the US – will always be here because much of the

Conclusion

land under cultivation is ideally suited to growing cereals. Universal self-sufficiency is an impossible goal in a world of climate insecurity. It is what cereals the major exporters grow and how they do it that need to change. Population cereals, mixtures that do best in low-input agroecological systems and that express great heterogeneity, improve soil health and biodiversity, and sequester carbon must win out over homogeneity, which achieves none of those things. I believe they can do it, because the model of homogeneity is unsustainable. Already the dominant players in plant breeding bandy around the language of the alternative, appropriating words like regenerative, agroecological, sustainable, to bend the narrative to their way of working. I see this linguistic sleight of hand as evidence that they know their argument is lost and they will have to change or perish.

So many countries in the Global South already have strategies for improved plant breeding that are inspired by indigenous farming knowledge and expertise. But this critical human capital is frequently expected to exchange practices that underpin resilience and adaptiveness for a Western, technological solution that is informed by a hangover of the colonial attitude that 'we know best'. An ambition to transform agriculture in Africa with biotechnology supported by hundreds of millions of dollars from the Bill and Melinda Gates Foundation into funding development of GMO seeds has been a notable failure.[4] Forced upon nations by governments enamoured of the sound of US money and the lure of collaboration between local and international research institutions, 60 per cent of publicly funded breeding projects have failed to get beyond the research phase; evidence, if it was needed, that money is being wasted on promoting a vision to improve African agriculture that is expensive and not fit

The Accidental Seed Heroes

for purpose. A top-down and one-size-fits-all approach that is dependent on patented seeds and offers no solutions that are truly adapted to local conditions does nothing to realise the amazing potential for the continent to be completely self-sufficient in food and able to counter climate extremes without being dependent on aid.

I saw for myself in Ethiopia the waste of human capital due to the lack of availability of water, with millions of citizens spending hours every day carrying water for their livestock and domestic use. But the pressure on land is huge, because so much that could be cultivated lacks water. Ethiopia is not short of it. Some of Africa's greatest rivers flow through the country and the Rift Valley is home to several huge lakes, fed from the high mountains above. When I spoke with agronomists and plant scientists at the Ethiopian Biodiversity Institute (EBI), the message was loud and clear: there are 38.5 million hectares of arable land in Ethiopia, but just 16.2 million – a mere 42 per cent – are under cultivation. Getting water to the fertile dry plains in the southeast, utilising a programme to lower salinity through agroecological means, can achieve a number of things. It would transform Ethiopia's food economy and feed much of Africa, and it would enable the growing population to have land on which to live and cultivate crops. Aid is always a last resort – providing food when everyone is starving. The folks at the EBI told me that investing in water infrastructure would make the World Food Programme in Ethiopia obsolete. Would that Bill Gates had put his foundation's money into such an ambition rather than supporting an unsustainable model of food production underpinned by a monopolistic mindset. It's not too late. He should get his cheque book out right now and do something that could really help.

Conclusion

Governments too can provide leadership and the framework for people-focused solutions and approaches to problem solving in breeding and adaptation. At the time of writing the UK, unlike most other major economies, does not have a national food policy. Nutritious, accessible and affordable food needs to be part of national strategy, not something that comes to the fore only when a country faces crises such as war, trade disruption or the politics of price. With a greater diversity in adaptive and locally bred varieties, moving away from homogeneity towards greater heterogeneity, nations will become more self-sufficient and that includes my own country.

We think of artists – painters, poets, composers, novelists and all those in the creative industries – as inventors and reflectors of culture. Through their tools – pencil, paint, mouse – we can place ourselves within a cultural space, feel an identity. And so it is with farmers. But rather than pencils and paint in their hands they hold seeds. As well as being the source of our physical nourishment, like a painting or a poem they feed our souls too. This reality is just one reason why farmer-centred breeding must be at the very core of our approach to feeding ourselves and at the same time nourishing our planet. I have borne witness to a counter-revolution. Like a hyperbolic curve with a long, slow initial climb up the graph, it's been gradually gaining strength and signs of increased momentum for at least the last quarter century. I have no doubt that future generations will look back on the twenty-first century as the period in which agricultural sanity was restored. Those of us who save seeds from year to year, regardless of how well organised and selective we are in keeping back the best from the crop, are all plant breeders. Our much-loved favourites and newly discovered harbingers

The Accidental Seed Heroes

of deliciousness are continuing to evolve. We accidentally create new FVs, locally adapted to do well in our gardens and allotments. From time to time, individual plants mutate or accidentally cross – and when it is in a good way, resulting in improved vigour, resilience and flavour, we have created a new variety. So, give it a name and make it a part of your life story. The importance of diversity in the home garden cannot be overestimated. Open-source plant breeding, as championed by the Open Source Seed Initiative in the US and increasingly around the world, empowers everyone, amateur and professional, who grows and saves seeds to be part of building resilience and sustainability into our future foods. I'm proud to have joined the party, adding to the genetic diversity of edible crops, which are such a big part of the solution to feeding ourselves in an ever more chaotic climate.

Acknowledgements

As a film producer, I always wanted to make programmes on subjects I knew nothing about. The success, or otherwise, of the many journeys of discovery I took to find and tell stories that assuaged my curiosity were only possible thanks to the help and support of people who knew considerably more about the subject and the places than I did. So, attempting to write a book about a pursuit for deliciousness that could also inform the existential question of how we develop crops that will feed the world in a time of climate change and the need to achieve net zero in all aspects of our lives by mid-century made me entirely dependent on the expertise of others.

First and foremost, I must thank my beloved wife, Julia, who read every word as the book took shape and asked enquiring and challenging questions, which, I hope, have helped make the work more readable. Her belief in me as a writer, putting up with my frequent absences travelling to inspiring countries, making sure I found the proper relationship between my backside and the chair in my study, made the task I set myself a pleasure.

This book would never have been written without the encouragement and support of Muna Reyal and the lovely folks at Chelsea Green, Matt Haslum, Rosie Baldwin, Alex Stewart and Caroline Taggart, whose consummate skills as an editor to help make this book better have made for a delightful collaboration. And thanks also to copy editor

Susan Pegg, whose forensic grappling with the text and sharp eye for a fact has kept me on my toes.

Thanks, too, for sage advice and endless encouragement from my agent, Sonia Land.

My heartfelt thanks go also to the many people – farmers, academics, professional and amateur plant breeders and assorted opinion formers and activists – who have answered my questions and read many chapters of the book with invaluable suggestion to improve my prose, get my facts straight and remind me that I have merely scratched the surface of this, to me at least, profoundly fascinating and inspiring subject.

Some of the names featured in these pages I want to extend special thanks and gratitude to are: C.R. Lawn, Alex McAlvay, Andrew Flachs and Steven Jacobs, who engaged in much enlightening discourse – for me, at least – read various drafts, and whose critical and forensic suggestions for improvements I could not have completed the work without.

I extend deep gratitude to the many others who took time to share their knowledge and provided much support and encouragement. From the Royal Botanic Gardens Kew: Philippa Ryan and James Borrell. From Albania: Lavdosh Ferruni, Robert Damo and Socrat Yani. From Denmark: Eva Skjødsholm, John Kelly, Søren Holt and Ærling Frederiksen. On the subject of cereals: Anders Borgen, Andrew Whitley and Anne Parry. For sharing their knowledge of Ethiopia: Zemede Asfaw, Melesse Maryo, Regassa Feyissa, Wubeshet Teshome, Leta Ajema, Basazen Fantahun and Tamene Yohannes. All things to do with legumes: Agnese Brantestam, Pete Iannetta and Josiah Meldrum. My discoveries in Austria: Josef Obermoser, Wolfgang Palme, Andreas Motschiunig, Anna Ambrosch, Peter Laßnig, Helene Maierhofer and all

Acknowledgements

the folks at Arche Noah. For their answers to my plant breeding questions: Katharine McEvoy, Bruce Pearce, Gabriella Morini, Grietje Raaphorst, Lieven David, Fred Groom and Ronja Schlumberger, Noel Ellis, Christopher Judge, Tom Hartley, Maren Korsgaard, John Bunker, Craig LeHoullier, Joy Michaud, Christopher Judge. And from the John Innes Centre: Sarah Wilmot, Simon Griffiths and Noel Ellis.

To everyone who has helped and shared in my journey, if I have got things wrong or not included you in my acknowledgements, I apologise. All errors of fact, botanical and historical rest entirely at my door.

But above all, it is the amazing and lovely farmers and growers that I have met on my travels I have to thank the most. It is in their hands that the future of our food depends. I am happy to reiterate: I sit at their feet. Their knowledge, relationship with what they grow, connection to the land and community are ultimate expressions of what it is to be human, to love the world and to want to see it continue to blossom.

Glossary

Abiotic stress: The stress experienced by plants from the effects of natural events; temperature, light levels, wind, drought, flood and salinity.

Accession: A distinct element of a collection of plant material from a single species, found in one location or acquired at the same time and representing the diversity present in a given population of plants.

Adapted: Describes a plant that has evolved or been bred to grow under certain measurable conditions, mostly associated with resistance to specific biotic and abiotic stresses and chemical and fertiliser regimes.

Adaptive: Describes a plant or population of plants that is genetically diverse and able to respond to changes in biotic and abiotic stresses over time.

Agroecology: An agricultural practice that is described by the Organisation for Economic Co-operation and Development (OECD) as 'the study of the relation of crops and the environment'. It is a holistic approach to producing food that brings together agriculture and community for the benefit of nature and economic wellbeing.

Biotic stress: A stress that occurs on and impacts plants as a result of damage by other living organisms such as bacteria, viruses, fungi, parasites, insects, both beneficial and harmful, and other plants, both wild and domesticated.

Glossary

Crop wild relative: A wild plant that is closely related to a domesticated one either as a direct ancestor of a cultivated species or another taxon.

Cultivar: A plant that is the result of direct intervention by human hand through selective breeding of known other cultivars or varieties and which has identifiable traits and characteristics including size, colour, shape and disease resistance. Most commercially grown varieties are cultivars.

Dehybridisation: The creation of new cultivars through an ongoing programme of selecting seed from the offspring of F1 hybrid cultivars over several generations.

Diploid: Describes a cell that contains two copies of each chromosome, one inherited from the male and the other from the female parent.

Distinct: Describes the unique traits expressed by a cultivar or variety.

Elite: A cultivar that has been bred for use in monocultures and intensive food production.

Evolutionary populations (EV): Also known as **bulk populations**, the result of **evolutionary plant breeding** whereby crops under cultivation evolve via natural selection pressures, adapting to a changing environment.

Flavonoid: A class of naturally occurring phenols and specialised metabolites that determine colour, nitrogen fixing and UV filtration in plants.

Food security: An increased level of self-sufficiency in national production, reducing dependence on imported and global food networks for a nation to feed itself.

FV (farmers' variety or folk variety): Also known as a landrace, a plant population with a limited range of genetic

The Accidental Seed Heroes

variations, which is adapted to local agroclimatic conditions and has been generated, selected, named and maintained by traditional and indigenous farmers.

Gene pool: All the genetic information and associated genes in a particular species.

Genetic modification (GM or GMO): An application of plant breeding in which DNA from one species is inserted into the genome of another.

Genetic variation: The difference in DNA between indivdual plants of a single variety or species or varietal populations; the result of mutations or genetic recombination.

Genome editing: A method whereby specialised enzymes are used to cut strings of DNA at certain points, which are equivalent to those changes that can occur using traditional breeding methods.

Genome sequencing: The process of determining all or part of the DNA sequence of a plant's genome.

Genotype: The complete set of genetic material that makes up an organism. The term also refers to the parts of the genetic makeup of a cell, alleles or variants a plant has in a particular gene or its location in the genome, which determine its characteristics.

Germplasm: The genetic resources, including seeds, tissue and DNA sequences, that are held by research institutions and plant breeders. These collections can include wild, elite and domesticated breeding lines that may have been reproduced for centuries and are critical in maintaining biological diversity, food security and conservation of genetic resources.

Haploid: Describes a cell that possesses one set of chromosomes that are present in eggs and male sperm cells.

Glossary

Heterogeneous: When a variety or species displays genetically determined variability in traits or attributes (such as flowering period, disease resistance, seed size, etc.), which might be of interest to plant breeders.

Heterosis (hybrid vigour): The result of selecting from inbred lines such that when crossed with a different inbred line the resulting plant expresses increased vigour in some aspects of its morphology.

Heterozygous: Refers to a pair of alleles within a gene that are different, one being inherited from the mother and the other from the father. In the laws of inheritance alleles are either dominant or recessive. It is the dominant allele that is expressed, although the recessive allele is still present and can be inherited by the next generation.

Hexaploid: Having six sets of chromosomes within the nucleus of an organism, as explored in this book with the bread wheat *Triticum aestivum*, which contains three closely related diploid genomes.

Homogeneous: Describes a crop where all the plants are genetically identical and derived from one common parent or parents.

Homozygous: Defines a specific gene where the two alleles are identical, either dominant (AA) or recessive (aa).

Inbred lines: Also known as **inbred strains**, individuals of a particular species or cultivar that have a nearly identical genotype due to at least twenty generations of brother/sister inbreeding.

Inbreeding depression: The loss of genetic diversity that results from inbreeding from small populations of plants. This often happens due to a dramatic reduction in size caused by biotic and abiotic stresses creating a **population bottleneck.**

Ideotype: Described by C.M. Donald in 1968 as an idealised structural model of a plant that can theoretically attain maximum yield quantity and vigour within a given environment.[1]

Introgression: the infiltration of genes from one population into another population of a different genotype.

Landrace: See **FV.**

Marker-assisted selection (MAS): Also known as **marker-aided selection,** this is a scientific process where a trait that is of interest to a plant breeder is selected from a biological marker in the genome of one species or variety that expresses the desired trait and can be employed through classical or molecular breeding strategies.

Maslin: A traditional cereal mix, grown across Africa and Eurasia since the Bronze Age. Unlike varietal mixtures such as wheat populations, maslins contain more than one species. Because of this, they are not thought of as populations from an ecological and evolutionary perspective, but rather as ecological communities.

Morphology: A branch of biology that deals with the study of the form and structures of an organism, including its size and shape.

Neuroactive compounds: These include serotonin and melatonin, which are chemicals synthesised by plants and micro-organisms, often as a stress reaction, that can affect human health and mood.

Organoleptic: Refers to qualities of a food that stimulate the senses, including taste, colour, smell, appearance and feel.

Parthenocarpy: The natural or induced development of a fruit without it being fertilised, first described in 1902

Glossary

by Fritz Noll.[2] It is employed in commercial breeding to develop seedless crops including aubergine, cucumbers and some citrus like tangerines, bananas and figs.

Pathogenicity: The absolute ability of an infectious agent (which may or may not be a **pathogen**) to cause disease or damage in its host plant.

Phenotype: The observable characteristics or traits of a species, including its physical form and structure (morphology).

Polyploid: Describes a species whose nuclei have more than two sets of chromosomes that are, in many instances, derived from different species.

Precision breeding: A technique that uses modern technologies like gene editing to change the DNA of a plant or animal in a precise way. Some argue that this dramatically reduces the time breeders need to develop new cultivars for a changing climate and more mouths to feed.

Regenerative agriculture/farming: A term often loosely used to describe an approach to cultivation that focuses on the conservation and restoration of biodiversity, soil health, carbon sequestration, water conservation and increased resilience to climate change, employing mostly but not exclusively organic principles.

Resilience: The scientific definition of resilience is the ability to bounce back after experiencing a shock. Applied to plants it means just that.

Taxon: A group of one or more populations of plants that taxonomically form a unit and are usually, but not always, given a name and ranking in the taxonomic hereditary hierarchy seen in plant classification.

Tetraploid: Describes a species whose nuclei have three sets of chromosomes of which at least one is from a different species.

The Accidental Seed Heroes

Tilling: A method in molecular biology that enables the identification of mutations within a specific gene and is used in developing new **genetically modified crops**, including maize, wheat, soya bean, tomato and lettuce.

Transgenic: Describes an organism (plant) that has had DNA from another species introduced into its genome through a process of genetic modification.

Variety: A naturally occurring subset within a species that shows clearly identifiable traits and characteristics, including shape, size, colour and disease resistance, and does not require human intervention to reproduce. Varieties are associated with heirloom and heritage crops that have not been deliberately bred.

Zygosity: The extent to which both copies of a gene or chromosome have the same genetic sequence. Low zygosity illustrates little or no difference, whereas at the other end of the spectrum extreme heterozygotes (such as apples) mean there is little or no similarity in the sequences.

Notes

Introduction

1. This is a must-read for anyone who wants to breed vegetables themselves: Carol Deppe, *Breed Your Own Vegetable Varieties: The Gardener's and Farmer's Guide to Plant Breeding and Seed Saving* (Chelsea Green Publishing, 2000).
2. More about the Open Source Seed Initiative and its work can be found here: https://osseeds.org.

Chapter 1. Breaking the Mould

1. Harley Harris Bartlett, 'The Reports of the Wilkes Expedition, and the Work of the Specialists in Science', *Proceedings of the American Philosophical Society* 82, no. 5 (1940): 675–78.
2. Susan E. Gustad, 'Legal Ownership of Plant Genetic Resources – Fewer Options for Farmers', *Hamline Law Review* 18, no. 3 (1994–1995): 464.
3. European Patent Convention Rule 26(4) defines 'plant variety' as meaning: 'any plant grouping within a single botanical taxon of the lowest rank, which grouping, irrespective of whether the conditions for the grant of a plant variety right are fully met, can be: (a) defined by the expression of the characteristics that results from a given genotype or combination of genotypes, (b) distinguished from any other plant grouping by the expression of at least one of the said characteristics, and (c) considered as a unit with regard to its suitability for being propagated or unchanged.', https://www.epo.org/en/legal/epc/2020/r26.html.

The Accidental Seed Heroes

4. More on this story can be found here: Denis Meshaka, 'A Dutch Seed Company Faces up to KWS Patents', Inf'OGM 4 April 2024, https://infogm.org/en/a-dutch-seed-company-faces-up-to-kws-patents.

5. For a compelling and detailed telling of the story of the patenting of seeds, it is worth reading Jack Ralph Kloppenburg, *First the Seed: The Political Economy of Plant Biotechnology*, 2nd ed. (University of Wisconsin Press, 2005).

6. Organisation for Economic Cooperation and Development, *Concentration in Seed Markets: Potential Effects and Policy Responses* (OECD Publishing, 2018), 163–77, https://doi.org/10.1787/9789264308367-en.

7. Sarah K. Lowder, Jakob Skoet and Saumya Singh, eds. 'What Do We Really Know about the Number and Distribution of Farms and Family Farms in the World? Background Paper for The State of Food and Agriculture 2014', ESA Working Paper 14-02 (2014), https://doi.org/10.22004/ag.econ.288983.

8. ETC Group, *Who Will Feed Us?*, 3rd ed. (ETC Group, 2017), https://www.etcgroup.org/whowillfeedus; A Growing Culture, 'Can Small-Scale Farmers Feed the World?', Offshoot, 2 August 2022, https://agrowingculture.substack.com/p/can-small-scale-farmers-feed-the.

9. Vincent Ricciardi et al., 'How Much of the World's Food Do Smallholders Produce?', *Global Food Security* 17 (2018): 64–72, https://doi.org/10.1016/j.gfs.2018.05.002; Sarah K. Lowder, Marco V. Sánchez and Raffaele Bertini, 'Which Farms Feed the World and Has Farmland Become More Concentrated?' *World Development* 142 (2021): 105455, https://doi.org/10.1016/j.worlddev.2021.105455.

10. Lowder, Sánchez and Bertini, 'Which Farms Feed the World', 105455.

11. Ricciardi et al., 'How Much of the World's Food Do Smallholders Produce?', 64–72; Leah H. Samberg et al., 'Subnational Distribution of Average Farm Size and Smallholder Contributions to Global Food Production', *Environmental Research Letters* 11, no. 12 (November 2016): 124010, https://doi.org/10.1088/1748-9326/11/12/124010.

12. A Growing Culture, 'Can Small-Scale Farmers Feed the World?', Offshoot, 2 August 2022, https://agrowingculture.substack.com/p/can-small-scale-farmers-feed-the.

13. A scholarly analysis of the state of IP in the US can be found here: Paulina B. Jenney, 'Keeping What You Sow: Intellectual Property Rights for Plant Breeders and Seed Growers', *Graduate Student Theses*,

Notes

Dissertations, and Professional Papers (2022): 11928, https://scholarworks
.umt.edu/etd/11928.

Chapter 2. Where Farmers' Varieties Reign Supreme

1. Sokrat Jani, Adriatik Cakalli, Valbona Hobdari and Damo Robert, 'Case Study of Local Varieties and Landraces of Vegetable Crops in Korça Region', *Albanian Journal of Agricultural Sciences* 19, no. 2 (2020): 31–37.
2. James S. Borrell et al., 'Enset in Ethiopia: A Poorly Characterized but Resilient Starch Staple', *Annals of Botany* 123, no. 5 (2019): 747–66, https://doi.org/10.1093/aob/mcy214.
3. A description of the salt mines of the Danakil Depression can be found here: Bob Koigi, 'The Hottest Place on Earth: The Salt Mines of Danakil', FairPlanet, https://www.fairplanet.org/story/the-hottest-place-on-earth -the-salt-mines-of-danakil-depression.
4. Tafesse Matewos, 'Climate Change-Induced Impacts on Smallholder Farmers in Selected Districts of Sidama, Southern Ethiopia'. *Climate* 7, no. 5 (May 2019): 70, https://doi.org/10.3390/cli7050070.
5. Meyer, Rachel S., Ashley E. DuVal and Helen R. Jensen. 'Patterns and Processes in Crop Domestication: An Historical Review and Quantitative Analysis of 203 Global Food Crops', *New Phytologist* 196, no. 1 (2012): 29–48, https://doi.org/10.1111/j.1469-8137.2012.04253.x.
6. Oliver W. White et al., 'Maintenance and Expansion of Genetic and Trait Variation Following Domestication in a Clonal Crop', *Molecular Ecology* 32, no. 15 (2023): 4165–80, https://doi.org/10.1111/mec.17033.
7. Rachel R. Chase et al., 'Smallholder Farmers Expand Production Area of the Perennial Crop Enset as a Climate Coping Strategy in a Drought-Prone Indigenous Agrisystem', *Plants, People, Planet* 5, no. 2 (2023): 254–66, https://doi.org/10.1002/ppp3.10339.
8. Sokrat, Cakalli, Hobdari and Robert, 'Case Study of Local Varieties and Landraces', 31–37.
9. Damo Robert, Pirro Icka and Hajri Haska, 'Traditional Local Varieties and Adaptation to Climate Change in Korça Region, Albania Varietetet Lokale Tradicionale Dhe Përshtatja Ndaj Ndyshimeve Klimatike Në Rajoninin e Korçës, Shqipëri', 1st International Conference on Plant Eco-physiological Adaptation Mechanisms - PEPAM, Tirana, Albania, 21–22 July, 2022, https://doi.org/10.13140/RG.2.2.18084.14721.

Chapter 3. Using One's Loaf

1. Jenkins, T. M. 'No. 1. Pure Lines of Hen Gymro Wheat'. *New Varieties and Strains from The Welsh Plant Breeding Station*, S, No. 1 (1928), http://www.brockwell-bake.org.uk/docs/No.1 per cent20Pure per cent20Lines per cent20of per cent20Hen per cent20Gymro per cent20Wheat_WPBS_OCR.pdf.

2. Iris Groman-Yaroslavski, Ehud Weiss and Dani Nadel. 'Composite Sickles and Cereal Harvesting Methods at 23,000-Years-Old Ohalo II, Israel'. *PLoS ONE* 11, no. 11 (23 November 2016): e0167151, https://doi.org/10.1371/journal.pone.0167151.

3. Gordon C. Hillman and M. Stuart Davies, 'Measured Domestication Rates in Wild Wheats and Barley Under Primitive Cultivation, and Their Archaeological Implications', *Journal of World Prehistory* 4, no. 2 (1990): 157–222.

4. Jizeng Jia et al., 'Aegilops Tauschii Draft Genome Sequence Reveals a Gene Repertoire for Wheat Adaptation', *Nature* 496, no. 7443 (April 2013): 91–95, https://doi.org/10.1038/nature12028.

5. Edward R. Treasure, Darren R. Gröcke, Astrid E. Caseldine and Mike J. Church, 'Neolithic Farming and Wild Plant Exploitation in Western Britain: Archaeobotanical and Crop Stable Isotope Evidence from Wales (c. 4000–2200 Cal Bc)', *Proceedings of the Prehistoric Society* 85 (December 2019): 193–222, https://doi.org/10.1017/ppr.2019.12.

6. Adnan Riaz et al., 'Mining Vavilov's Treasure Chest of Wheat Diversity for Adult Plant Resistance to Puccinia Triticina', *Plant Disease* 101, no. 2 (February 2017): 317–23, https://doi.org/10.1094/PDIS-05-16-0614-RE.

7. The Brockwell Bake Association, 'Hen Gymro', Wheat Database Gateway Portal, 31 July 2014, https://www.wheat-gateway.org.uk/hub.php?ID=41.

8. 'Andrew Forbes, Brockwell Bake Association, https://brockwell-bake.academia.edu/AndrewForbes; The work of Brockwell Bake association can be found here: http://www.brockwell-bake.org.uk.

9. Ambrogio Costanzo et al., 'Milestone: No18, M2.8 Delivery Date: 28 May 2018 Status: Approved Version 1.3', n.d.

10. Dorian Q. Fuller and Chris J. Stevens, 'Sorghum Domestication and Diversification: A Current Archaeobotanical Perspective', in *Plants and People in the African Past: Progress in African Archaeobotany*, ed. Anna Maria Mercuri, A. Catherine D'Andrea, Rita Fornaciari and Alexa Höhn

Notes

(Springer International Publishing, 2018): 427–52, https://doi.org /10.1007/978-3-319-89839-1_19.

11. A. Teshome et al., 'Maintenance of Sorghum (Sorghum Bicolor, Poaceae) Landrace Diversity by Farmers' Selection in Ethiopia', *Economic Botany* 53, no. 1 (1 January 1999): 79–88, https://doi.org/10.1007/BF02860796.

12. Jeff Mulhollem, 'Flavonoids' Presence in Sorghum Roots May Lead to Frost-Resistant Crop', Penn State University', 10 August 2020, https:// www.psu.edu/news/research/story/flavonoids-presence-sorghum-roots -may-lead-frost-resistant-crop.

13. Dorothy A. Mbuvi et al., 'Novel Sources of Witchweed (Striga) Resistance from Wild Sorghum Accessions', *Frontiers in Plant Science* 8 (2017), https://doi.org/10.3389/fpls.2017.00116.

14. Mbuvi et al., 'Novel Sources of Witchweed'.

15. Alex C. McAlvay et al., 'Cereal Species Mixtures: An Ancient Practice with Potential for Climate Resilience. A Review', *Agronomy for Sustainable Development* 42, no. 5 (13 October 2022): 100, https://doi.org/10.1007 /s13593-022-00832-1.

16. Lauren D. Snyder, Miguel I. Gómez and Alison G. Power, 'Crop Varietal Mixtures as a Strategy to Support Insect Pest Control, Yield, Economic, and Nutritional Services', *Frontiers in Sustainable Food Systems* 4 (2020), https://doi.org/10.3389/fsufs.2020.00060.

17. Susan Kelly, 'Ancient Farming Strategy Holds Promise for Climate Resilience', Cornell Chronicle, https://news.cornell.edu/stories/2023/01 /ancient-farming-strategy-holds-promise-climate-resilience.

Chapter 4. Setting Seeds Free

1. Alyssa N Crittenden and Stephanie L. Schnorr, 'Current Views on Hunter-Gatherer Nutrition and the Evolution of the Human Diet', *American Journal of Physical Anthropology* 162, no. S63 (2017): e23148, https://doi .org/10.1002/ajpa.23148.

2. Peter Frankopan, *The Earth Transformed* (Bloomsbury, 2007), 108–110.

3. John Warren, *The Nature of Crops: How We Came to Eat the Plants We Do* (CABI Publishing, 2015).

4. Rafa Ramos, 'Statement on the Dakar 2 Summit : "Climate Smart Agriculture" Will Worsen the Climate Crisis', *Assess Technology*,

25 January 2023, https://assess.technology/featured/statement
-dakar-summit-climate-smart-agriculture-climate-crisis.

5. Helen Anne Curry, 'Hybrid Seeds in History and Historiography', *Isis*
113, no. 3 (2022): 610–17.

6. Jonathan Foley, 'It's Time to Rethink America's Corn System', *Scientific
American*, 5 March 2013, https://www.scientificamerican.com/article
/time-to-rethink-corn.

7. Jonas Jägermeyr et al., 'Climate Impacts on Global Agriculture Emerge
Earlier in New Generation of Climate and Crop Models', *Nature Food* 2,
no. 11 (November 2021): 873–85, https://doi.org/10.1038/s43016-021
-00400-y; Tianyi Zhang et al., 'Increased Wheat Price Spikes and Larger
Economic Inequality with 2°C Global Warming', *One Earth* 5, no. 8
(19 August 2022): 907–16, https://doi.org/10.1016/j.oneear.2022.07.004.

8. Mudigere Sannegowda, Umesh Babu and Satish Chandra Garkoti,
'Traditional Community-Led Seed System for Maintaining Crop Vigour,
Diversity and Socio-Cultural Network in View of the Changing Climate:
A Case Study from Western Himalaya, India', *Climate Action* 1, no. 1
(17 August 2022): 1–16, https://doi.org/10.1007/s44168-022-00020-7.

9. Bharat Ramaswami, Carl E. Pray and N. Lalitha, 'The Spread of Illegal
Transgenic Cotton Varieties in India: Biosafety Regulation, Monopoly,
and Enforcement', *World Development* 40, no. 1 (1 January 2012): 177–88,
https://doi.org/10.1016/j.worlddev.2011.04.007.

10. Salvatore Ceccarelli, Stefania Grando, Maedeh Salimi and Khadija Razavi,
'Evolutionary Populations for Sustainable Food Security and Food
Sovereignty', in *Seeds for Diversity and Inclusion: Agroecology and Endogen-
ous Development*, ed. Yoshiaki Nishikawa and Michel Pimbert (Springer
International Publishing, 2022), https://doi.org/10.1007/978-3-030
-89405-4_8.

11. J.B. Smithson and J M. Lenné, 'Varietal Mixtures: A Viable Strategy for
Sustainable Productivity in Subsistence Agriculture', *Annals of Applied
Biology* 128, no. 1 (1996): 127–58, https://doi.org/10.1111/j.1744-7348
.1996.tb07096.x.

12. R.W. Allard, 'Relationship Between Genetic Diversity and Consistency
of Performance in Different Environments1', *Crop Science* 1, no. 2 (1961):
cropsci1961.0011183X000100020012x, https://doi.org/10.2135
/cropsci1961.0011183X000100020012x.

Notes

13. Bocci, Riccardo et al., 'Yield, Yield Stability and Farmers' Preferences of Evolutionary Populations of Bread Wheat: A Dynamic Solution to Climate Change', *European Journal of Agronomy* 121 (2020): 126156, https://doi.org/10.1016/j.eja.2020.126156; Isabelle Goldringer et al., 'Rapid Differentiation of Experimental Populations of Wheat for Heading Time in Response to Local Climatic Conditions', *Annals of Botany* 98, no. 4 (2006): 805–17, https://doi.org/10.1093/aob/mcl160.
14. Salvatore Ceccarelli and Stefania Grando, 'Evolutionary Plant Breeding as a Response to the Complexity of Climate Change', *iScience* 23, no. 12 (18 December 2020): 101815, https://doi.org/10.1007/978-3-030-89405-4_8.
15. Ernest Small, '47. Teff & Fonio – Africa's Sustainable Cereals', *Biodiversity* 16, no. 1 (January 2, 2015): 27–41, https://doi.org/10.1080/14888386.2014.997290.
16. Tadesse Ebba, 'T'éf (*Eragrostis tef*): The Cultivation, Usage, and Some of the Known Diseases and Insect Pests', College of Agriculture Haile Selassie I University, 1969.
17. K. Assefa et al., 'Breeding Tef [Eragrostis Tef (Zucc.) Trotter]: Conventional and Molecular Approaches', *Plant Breeding* 130, no. 1 (2011): 1–9, https://doi.org/10.1111/j.1439-0523.2010.01782.x.
18. K. Assefa, 'Breeding Tef', 1–9.
19. H.K. Jifar, Tesfaye, K. Assefa, S. Chanyalew and Z. Tadele. 'Semi-Dwarf Tef Lines for High Seed Yield and Lodging Tolerance in Central Ethiopia', *African Crop Science Journal* 25, no. 4 (27 November 2017): 419–39, https://doi.org/10.4314/acsj.v25i4.3.
20. Details of this collaboration can be found here: 'Tef Improvement Project', University of Bern, https://www.tef-research.org/index.html.
21. Gina Cannarozzi et al., 'Technology Generation to Dissemination: Lessons Learned from the Tef Improvement Project" *Euphytica* 214, no. 2 (February 2018): 31, https://doi.org/10.1007/s10681-018-2115-5.

Chapter 5. A Future Full of Beans

1. 'Catalonian PDO Legumes: A Must-Try Culinary Gem', Foods and Wines from Spain, 23 February 2023, https://www.foodswinesfromspain.com/en/food/articles/2023/february/catalonian-pdo-legumes--a-culinary-gem.
2. 'These Are the 29 Facts You Never Knew about Baked Beans', Iceni Magazine, 23 October 2017, https://www.icenimagazine.co.uk/29-facts-never-knew-baked-beans.

The Accidental Seed Heroes

3. 'Farm Update 205 Princes Canning Capuat, A World First For A British Bred Baked Bean, 2023', posted by WardysWaffle Andrew Ward Farmer, 5 December 2023, https://www.youtube.com/watch?v=1SY0EO1SzGI.

4. Chhana Ullah et al., 'Flavan-3-Ols Are an Effective Chemical Defense against Rust Infection1[OPEN]', *Plant Physiology* 175, no. 4 (December 2017): 1560–78, https://doi.org/10.1104/pp.17.00842.

5. John A. Schnittker, 'The 1972-73 Food Price Spiral', in *Brookings Papers on Economic Activity*, 498–507, 1973, https://www.brookings.edu/wp-content/uploads/1973/06/1973b_bpea_schnittker.pdf.

6. D.A. Bond, 'Prospects for Commercialisation of F1 Hybrid Field Beans Vicia Faba L', *Euphytica* 41, no. 1 (1 April 1989): 81–86, https://doi.org/10.1007/BF00022415.

7. Kedar N. Adhikari et al., 'Conventional and Molecular Breeding Tools for Accelerating Genetic Gain in Faba Bean (Vicia Faba L.)', *Frontiers in Plant Science* 12 (2021), https://www.frontiersin.org/articles/10.3389/fpls.2021.744259.

8. James B. Manson et al., 'Genetic Gain in Yield of Australian Faba Bean since 1980 and Associated Shifts in the Phenotype: Growth, Partitioning, Phenology, and Resistance to Lodging and Disease', *Field Crops Research* 318 (1 November 2024): 109575, https://doi.org/10.1016/j.fcr.2024.109575; M.M.F. Abdalla et al., 'Performance and Reaction of Faba Bean Genotypes to Chocolate Spot Disease', *Bulletin of the National Research Centre* 45, no. 1 (17 September 2021): 154, https://doi.org/10.1186/s42269-021-00614-x.

9. Usman Ijaz et al., 'Mapping of Two New Rust Resistance Genes Uvf-2 and Uvf-3 in Faba Bean', *Agronomy* 11, no. 7 (July 2021): 1370, https://doi.org/10.3390/agronomy11071370.

10. 'Shaping the Future of Organic Breeding and Farming', BRESOV, 19 June 2018, https://www.bresov.eu.

11. 'BRESOV Factsheet', BRESOV, 24 June 2021, https://www.bresov.eu/about/_background/BRESOV_Factsheet.pdf.

12. P.B. Manjunatha et al., 'Exploring the World of Mungbean: Uncovering its Origins, Taxonomy, Genetic Resources and Research Approaches', *International Journal of Plant and Soil Science* 35, no. 20 (2023): 614-35, https://doi.org/10.9734/ijpss/2023/v35i203846.

Notes

13. Valérie Heuzé, Gilles Tran, Denis Bastianelli and François Lebas, 'Mung Bean (Vigna radiata)', Feedipedia, a programme by INRAE, CIRAD, AFZ and FAO, last modified 3 July 2015, https://www.feedipedia.org/node/235.
14. C. J. Lambrides and I.D. Godwin, 'Mungbean', in *Pulses, Sugar and Tuber Crops*, ed. Chittaranjan Kole (Springer, 2007), 69–90, https://doi.org /10.1007/978-3-540-34516-9_4.
15. Donna M. Winham and Andrea M. Hutchins, 'Perceptions of Flatulence from Bean Consumption Among Adults in 3 Feeding Studies', *Nutrition Journal* 10, no. 1 (21 November 2011): 128, https://doi.org/10.1186/1475 -2891-10-128.

Chapter 6. Cultivating Capsicums

1. Sagimbayev Adilgali and Sarkenayeva Ayazhan, 'Kokzhar Fair: History And Modernity', *Economics and Management* 16, no. 2 (2019): 200–211, https://ideas.repec.org/a/neo/journl/v16y2019i2p200-211.html.
2. 'Top 10+ Largest Producers Of Chilli', FrutPlanet, 12 November 2023, https://frutplanet.com/largest-producers-of-chilli-today.
3. K.M. Rezaul Karim et al., 'Current and Prospective Strategies in the Varietal Improvement of Chilli (*Capsicum annuum L.*) Specially Heterosis Breeding', *Agronomy* 11, no. 11 (November 2021): 2217, https://doi .org/10.3390/agronomy11112217.
4. Rajesh S. Patil, 'Conventional Plant Breeding v/s MAS, Which One Is Better', ResearchGate Discussion, https://www.researchgate.net/post /Conventional_plant_breeding_v_s_MAS_which_one_is_better.
5. Bartholomeus Pasangka and Abdul Wahid, 'Genetic Engineering of Local Cayenne Pepper (*Capsicum frustescens L.*) Through Breeding With Multigamma Irradiation Methods to Obtain Superior Offspring', *Journal of Agricultural Science* 13, no. 12 (15 November 2021): 81, https://doi.org /10.5539/jas.v13n12p81.

Chapter 7. Red Is Not the Only Colour

1. Linda Calvin and Roberta Cook, 'North American Greenhouse Tomatoes Emerge as a Major Market Force', Economic Research Service USDA, 1 April 2005, https://www.ers.usda.gov/amber-waves/2005/april/north -american-greenhouse-tomatoes-emerge-as-a-major-market-force.

The Accidental Seed Heroes

2. Simone Perna et al., 'Bioactives Profile of Purple and Black Tomato: Potential Applications in the Pharmaceutical Field Purple and Black Tomato', *Indian Journal of Pharmaceutical Sciences* 82, no. 6 (31 December 2020): 928–34, https://doi.org/10.36468/pharmaceutical-sciences.724.

3. François-Xavier Branthôme, 'Worldwide (Total Fresh) Tomato Production in 2021', Tomato News, 23 February 2023, https://www.tomatonews.com/en/worldwide-total-fresh-tomato-production-in-2021_2_1911.html.

4. Roberta Cook, and Linda Calvin, 'Greenhouse Tomatoes Change the Dynamics of the North American Fresh Tomato Industry', Economic Research Report Number 2, USDA, April 2005, http://www.ers.usda.gov/publications/pub-details/?pubid=45477.

Chapter 8. Perfecting the Perfect Pea

1. Martyn Rix, Mark Nesbitt and Christabel King, '1063. *Lathyrus oleraceus* Lam.: Leguminosae', *Curtis's Botanical Magazine* 40, no. 2 (2023): 197–205, https://doi.org/10.1111/curt.12513.

2. A detailed account of the discovery of this landrace can be found here: https://www.impecta.se/sv/halsa-och-gladje/pa-spaning-efter-jattearten.

3. 'Gallatin Valley Seed Company History', Gallatin Valley Seed Company, 11 March 2009, https://gallatinvalleyseed.com/history.php.

4. Oscar Checa, Marino Rodriguez, Xingbo Wu and Matthew Blair, 'Introgression of the *Afila* Gene into Climbing Garden Pea (*Pisum sativum* L.)', *Agronomy* 10, no. 10 (10 October 2020): 1537, https://doi.org/10.3390/agronomy10101537.

5. V.K. Solovieva, 'New Forms of Vegetable Peas', *Agorbiologica* 5 (1958): 124–26.

6. Lukas Wille et al., 'Untangling the Pea Root Rot Complex Reveals Microbial Markers for Plant Health', *Frontiers in Plant Science* 12 (2021), https://doi.org/10.3389/fpls.2021.737820.

7. Naglaa A.S. Muhanna, Safa E. Elwan and Nsreen D. Dib, 'Biological Control of Root Rot Complex of Pea (*Pisum sativum* L.)', *Egyptian Journal of Phytopathology* 46, no. 1 (15 June 2018): 49–67, https://doi.org/10.21608/ejp.2018.87766.

8. Cemile Yılmaz and Vural Gökmen, 'Neuroactive Compounds in Foods: Occurrence, Mechanism and Potential Health Effects', *Food Research*

Notes

International 128 (February 2020): 108744 https://doi.org/10.1016/j
.foodres.2019.108744.

9. Anne Edwards et al., 'Genomics and Biochemical Analyses Reveal a
   Metabolon Key to β-L-ODAP Biosynthesis in *Lathyrus sativus*', *Nature
   Communications* 14, no. 1 (16 February 2023): 876, https://doi.org
   /10.1038/s41467-023-36503-2.
10. 'How the Plant with a Toxic Past Can Become a Climate-Smart Crop of
    Tomorrow', John Innes Centre, 10 July 2023, https://www.jic.ac.uk
    /press-release/how-the-plant-with-a-toxic-past-can-become-a-climate
    -smart-crop-of-tomorrow.

Chapter 9. Let Us Eat Leaves

1. Anqi Zhao, Elizabeth H. Jeffery and Michael J. Miller, 'Is Bitterness Only
   a Taste? The Expanding Area of Health Benefits of Brassica Vegetables
   and Potential for Bitter Taste Receptors to Support Health Benefits',
   Nutrients 14, no. 7 (January 2022): 1434, https://doi.org/10.3390
   /nu14071434.
2. Anne Cathrine Thorup et al., 'Strong and Bitter Vegetables from Traditional
   Cultivars and Cropping Methods Improve the Health Status of Type 2
   Diabetics: A Randomized Control Trial', *Nutrients* 13, no. 6 (June 2021):
   1813, https://doi.org/10.3390/nu13061813.
3. A.D. Dayan, 'What Killed Socrates? Toxicological Considerations and
   Questions', *Postgraduate Medical Journal* 85, no. 999 (1 January 2009):
   34–37, https://doi.org/10.1136/pgmj.2008.074922.
4. U. Nagaharu, 'Genome Analysis in Brassica with Special Reference to the
   Experimental Formation of *B. Napus* and Peculiar Mode of Fertilization',
   Japanese Journal of Botany 7, no. 7 (1935): 389–452.
5. Alison Tindale, 'Nine Star Perennial Broccoli', *The Backyard Larder* (blog),
   22 April 2017, https://backyardlarder.co.uk/2017/04/nine-star
   -perennial-broccoli.
6. Magdalena Achrem, Edyta Stępień and Anna Kalinka, 'Epigenetic Changes
   Occurring in Plant Inbreeding', *International Journal of Molecular Sciences*
   24, no. 6 (12 March 2023): 5407, https://doi.org/10.3390/ijms24065407.
7. Aurelio Bianchedi, 'I Radicchi Di Treviso: Storia, Coltivazione, Forzatura e
   Commercio', *L'Italia Agricola* 98 (1961): 37–51.

The Accidental Seed Heroes

8. Gianni Barcaccia, Andrea Ghedina and Margherita Lucchin, 'Current Advances in Genomics and Breeding of Leaf Chicory (*Cichorium intybus* L.)', *Agriculture* 6, no. 4 (2016): 50, https://doi.org/10.3390/agriculture6040050.

9. Margherita Lucchin, Serena Varotto, Gianni Barcaccia and Paolo Parrini, 'Chicory and Endive', in *Vegetables I: Asteraceae, Brassicaceae, Chenopodica-ceae, and Cucurbitaceae*, ed. Jaime Prohens and Fernando Nuez (Springer, 2008): 3–48, https://doi.org/10.1007/978-0-387-30443-4_1.

10. Muhammad Saeed et al., 'Chicory (*Cichorium intybus*) Herb: Chemical Composition, Pharmacology, Nutritional and Healthical Applications', *International Journal of Pharmacology* 13, no. 4 (2017): 351–60, https://doi.org/10.3923/ijp.2017.351.360.

11. A.J.M. Baker, R.D. Reeves and A.S.M. Hajar, 'Heavy Metal Accumulation and Tolerance in British Populations of the Metallophyte *Thlaspi caerulescens* J. & C. Presl (Brassicaceae)', *New Phytologist* 127, no. 1 (1994): 61–68, https://doi.org/10.1111/j.1469-8137.1994.tb04259.x.

12. Guangdi Li and Peter D. Kemp, 'Forage Chicory (*Cichorium intybus* L.): A Review of Its Agronomy and Animal Production', in *Advances in Agronomy*, ed. Donald L. Sparks (Academic Press, 2005), 88:187–222, https://doi.org/10.1016/S0065-2113(05)88005-8.

13. Roshan Kumar et al., 'Molecular Basis of the Evolution of Methylthioalkylmalate Synthase and the Diversity of Methionine-Derived Glucosinolates', *The Plant Cell* 31, no. 7 (July 2019): 1633–47, https://doi.org/10.1105/tpc.19.00046.

14. 'Karen's Fave', Wild Garden Seed, 11 January 2024, https://www.wildgardenseed.com/product_info.php?products_id=770.

15. Marin Scotten, '"Laying Claim to Nature's Work": Plant Patents Sow Fear Among Small Growers', *Guardian*, 25 January 2024, https://www.theguardian.com/environment/2024/jan/25/plant-patents-large-companies-intellectual-property-small-breeders.

Chapter 10. Beautiful Brinjal

1. Prabir Purkayahstha and Satyajit Rath, 'Bt Brinjal: Need to Refocus the Debate', *Economic and Political Weekly* 45, no. 20 (15 May 2010): 42–8, https://www.epw.in/journal/2010/20/perspectives/bt-brinjal-need-refocus-debate.html.

Notes

2. Monika Mandal and IndiaSpend.com, 'When India Already Grows Enough Brinjal, Why Does It Need a Genetically Modified Variety?', Scroll.in, 17 October 2020, https://scroll.in/article/975600/whenindia-already-grows-enough-brinjal-why-does-it-need-a-genetically-modified-variety.
3. Shanshan Zhou et al., 'The Molecular Mechanism of Eggplant Parthenocarpy Revealed Through a Combined Analysis of the Transcriptome and the Metabolome', *Industrial Crops and Products* 193 (1 March 2023): 116168, https://doi.org/10.1016/j.indcrop.2022.116168.
4. Andrew Flachs, 'Transgenic Cotton: High Hopes and Farming Reality', *Nature Plants* 3 (January 2017): 16212, https://doi.org/10.1038/nplants.2016.212.
5. Yanhui Lu et al., 'Widespread Adoption of Bt Cotton and Insecticide Decrease Promotes Biocontrol Services', *Nature* 487 (July 2012): 362–65, https://doi.org/10.1038/nature11153.
6. K.R. Kranthi and Glenn Davis Stone, 'Long-Term Impacts of Bt Cotton in India', *Nature Plants* 6, no. 3 (March 2020): 188–96, https://doi.org/10.1038/s41477-020-0615-5.
7. Vandana Shiva and Afsar Jafri, 'Failure of GMOs in India', We Are Greens, 22 December 2003, http://www.greens.org/s-r/33/33-04.html.
8. Anthony M. Shelton, 'Bt Eggplant: A Personal Account of Using Biotechnology to Improve the Lives of Resource-Poor Farmers', *American Entomologist* 67, no. 3 (Fall 2021): 52–9, https://doi.org/10.1093/ae/tmab036.
9. Joan Conrow, 'GMO Eggplant Is Documented Win for Resource-Poor Farmers', Alliance for Science, 16 September 2021, https://allianceforscience.org/blog/2021/09/gmo-eggplant-is-documented-win-for-resource-poor-farmers.
10. Anthony M. Shelton et al., 'Impact of Bt Brinjal Cultivation in the Market Value Chain in Five Districts of Bangladesh', *Frontiers in Bioengineering and Biotechnology* 8 (May 2020): 498, https://doi.org/10.3389/fbioe.2020.00498.

Chapter 11. It's All in the Pip

1. Dominique A.M. Noiton and Peter A. Alspach, 'Founding Clones, Inbreeding, Coancestry, and Status Number of Modern Apple Cultivars', *Journal of the American Society for Horticultural Science* 121, no. 5 (1 September 1996): 773–82, https://doi.org/10.21273/JASHS.121.5.773.

The Accidental Seed Heroes

2. A delightful personal account of the acquisition of Golden Delicious by Stark Brothers Nurseries can be found here: Stark Bro's Nurseries & Orchards Co., 'The Trail of The Golden Apple', 28 July 2016, https://www.starkbros.com/growing-guide/article/trail-of-the-golden-apple.

3. Adrian Higgins, 'Why the Red Delicious No Longer Is', *Washington Post*, 5 August 2005, http://www.washingtonpost.com/wp-dyn/content/article/2005/08/04/AR2005080402194.html.

4. Timothy Egan, '"Perfect" Apple Pushed Growers Into Debt', *The New York Times*, 24 August 2018, https://www.nytimes.com/2000/11/04/us/perfect-apple-pushed-growers-into-debt.html.

5. Fondazione Edmund Mach FEM, 'An Italian-Led International Research Consortium Decodes the Apple Genome', AlphaGalileo, 5 September 2010, https://web.archive.org/web/20100905085301/http://www.alphagalileo.org/ViewItem.aspx?ItemId=83717&CultureCode=en.

6. An excellent article about the state of the modern apple industry in the UK can be found here: 'Pink Lady v the British Apple', BBC News, 21 October 2013, https://www.bbc.com/news/magazine-24578762.

7. 'Apple', National Horticulture Board, 16 November 2006, https://nhb.gov.in/report_files/apple/APPLE.htm.

8. Naibin Duan et al., 'Genome Re-Sequencing Reveals the History of Apple and Supports a Two-Stage Model for Fruit Enlargement', *Nature Communications* 8, no. 1 (15 August 2017): 249, https://doi.org/10.1038/s41467-017-00336-7.

9. Hélène Muranty et al., 'Using Whole-Genome SNP Data to Reconstruct a Large Multi-Generation Pedigree in Apple Germplasm', *BMC Plant Biology* 20 (2 January 2020): 2, https://doi.org/10.1186/s12870-019-2171-6.

10. Kathrin Strohm, 'Of the 30,000 Apple Varieties Found All over the World Only 30 Are Used and Traded Commercially', Agri Benchmark, 19 June 2013, http://www.agribenchmark.org/agri-benchmark/did-you-know/einzelansicht/artikel//only-5500-wi.html.

11. Noiton and Alspach, 'Founding Clones', 773–82.

12. 'Forbidden Fruit: The Dramatic Rise in Dangerous Pesticides Found on Fruits and Vegetables Sold in Europe and Evidence That Governments Are Failing Their Legal Obligations', Pesticide Action Network Europe, 24 May 2022, https://www.pan-europe.info/resources/reports/2022/05/forbidden-fruit-dramatic-rise-dangerous-pesticides-found-fruits-and.

Notes

13. A.G. Brown, 'The Effect of Inbreeding on Vigour and Length of Juvenile Period in Apples', in *Proceedings of Eucarpia Fruit Section Symposium V: Top Fruit Breeding*, ed. A.G. Brown, R. Watkins and F. Alston (1973), 30-9.

14. Ayesha Tandon, 'Analysis: How UK Winters Are Getting Warmer and Wetter', Carbon Brief, 16 February 2024, https://www.carbonbrief.org /analysis-how-uk-winters-are-getting-warmer-and-wetter.

15. Carlota González Noguer, Alvaro Delgado, Mark Else and Paul Hadley, 'Apple (*Malus × domestica Borkh.*) Dormancy – a Review of Regulatory Mechanisms and Agroclimatic Requirements', *Frontiers in Horticulture* 2 (20 July 2023): 1217689, https://doi.org/10.3389/fhort.2023.1217689.

16. Maren Korsgaard and Torben Toldam-Andersen, 'Participative Breeding of Unique Danish Apple Cultivars', in *Proceedings of the 21st International Conference on Organic Fruit-Growing: Filderstadt 19–21 February 2024*, ed. FOEKO e.V. (2024), 19–23, https://researchprofiles.ku.dk/da /publications/deltager-baseret-for%C3%A6dling-af-unikke-danske -%C3%A6blesorter.

17. To learn more about this remarkable initiative, visit: www.gnarlypippins.com.

18. Sarah Kostick, 'Development of New Commercial Apple Cultivars', moderated by Sadie Barrett, virtual presentation, 20 March 2024, posted 26 March 2024 by IdahoComm, YouTube, 1.29.23, https://www.youtube .com/watch?v=JZsF5nyw2_M&list=PL0E60BSiNJBGbuDym XRJMBIm5EcliPAyF&index=3.

19. 'Afalau Cymru/ Apples of Wales', Carwyn Graves, 10 January 2020, https://carwyngraves.com/afalau-cymru-apples-of-wales.

20. B.T.P. Barker, 'Long Ashton Research Station, 1903–1953', *Journal of Horticultural Science* 28, no. 3 (1953): 149–51, https://doi.org/10.1080 /00221589.1953.11513779.

21. Helen Harper et al., 'The Long Ashton Legacy: Characterising United Kingdom West Country Cider Apples Using a Genotyping by Targeted Sequencing Approach', *Plants, People, Planet* 2, no. 2 (March 2020): 167–75, https://doi.org/10.1002/ppp3.10074.

Conclusion: Holding Truth to Power

1. Shifeng Cheng et al., 'Harnessing Landrace Diversity Empowers Wheat Breeding', *Nature* 632 (June 2024): 823–31, https://doi.org/10.1038 /s41586-024-07682-9.

2. The work of No Patents on Seed can be found here: https://www.no-patents-on-seeds.org/index.php/en. And their member organisations here: https://www.no-patents-on-seeds.org/index.php/en/about-us/member-organisations.
3. Natalie G. Mueller and Andrew Flachs, 'Domestication, Crop Breeding, and Genetic Modification Are Fundamentally Different Processes: Implications for Seed Sovereignty and Agrobiodiversity', *Agriculture and Human Values* 39, no. 1 (March 2022): 455–72, https://doi.org/10.1007/s10460-021-10265-3.
4. 'AGRA's Policy Influence Exposed', Alliance for Food Sovereignty in Africa, 28 August 2024, https://afsafrica.org/press-release-agras-policy-influence-exposed.

Glossary

1. C.M. Donald, 'The Breeding of Crop Ideotypes', *Euphytica* 17, no. 3 (December 1968): 385–403, https://doi.org/10.1007/BF00056241.
2. Fritz Noll, *Über Fruchtbildung Ohne Vorausgegangene Bestäubung (Parthenocarpie) Bei Der Gurke* (Universitäts-Buchdruckerei, 1902).

Index

Page numbers for glossary entries are in **bold**; page numbers followed by *n* indicate footnotes.

A

Abhilash tomato 133–4
abiotic stresses 62, 62*n*, 71, **248**
Aborti chickpea 39
Abundance oat 12–13
Abyssinian pea (*Pisum sativum* ssp. *abyssinicum*) 171–2, 171*n*
accession 35, **248**
adaptation strategy 86–7
adapted **248**
adaptive **248**
Addis Ababa University 45
Æliger Ælings Ærgte Ært (Honest Ærling's Genuine Eating Pea) 151
afila (Af) gene mutation 159, 160–1
Africa
  failed biotechnology 241–2
  see also Ethiopia
African milk tree (*Euphorbia trigona*) 76
agribusinesses / big seed companies 52, 106, 226, 240
  abuse of patent law 23–4, 191–3, 238–9
  alliance with governments 88
  business model 4, 6, 22–3, 27, 29, 30, 84
  commodification of seeds 34
  and control of seeds as intellectual property 4, 22–3, 25, 26–8, 30, 84
  and hybrid cultivar dependence 49
  moving power away from 237–8
  perpetuation of monocultures 6, 30, 206
  power and market domination 6, 26
  standing up to 82–5
Agricultural Research Council (ARC), Bristol 230
Agricultural Research Station, Bikaner 134
Agrii seed company 100, 101
agritourism 49, 50
Agroecological Living Landscapes (ALLS), Andhra Pradesh 235–6
agroecology 237, **248**
agroforestry 1–2, 42, 75–6
Albania 31
  beans 93–5
  collapse of Communism 35, 38, 47
  farmers under attack 46–8
  future developments 49
  pepper growing 114–19
  saving FVs 34–8
  seed savers 40–2
Albanian Institute of Plant Genetic Resources 35
Alexander, Jake 122
All American Selection (1977) 158
alleles 100
Almaz (Black Diamond) aubergine 198–9

271

The Accidental Seed Heroes

Ambrosch, Anna 138, 141–2, 145, 154–5
American Civil War 15
amicho 43
Andersen, Torben Bo Toldam 220–1
Andhra Pradesh 235–6, 237
Annie Elizabeth apple 225
Annual Wild and Seedling Pomological
  Exhibition, US 224
anthocyanins 141
antioxidants 104
apple forests 220–2
Apple Oasis, Denmark 220
Applegreen aubergine 198
apples (*Malus domestica*) 9, 211–32
  chill hour requirements 219–20, 228
  cider 223–4, 225, 227, 229–31
  founding clones 218
  grafting 216–17
  inbreeding 218–19
  new breeds 214–15
  self-seeded 222–3, 224
  twentieth-century superstars 212–14
  varieties vs cultivars 218*n*
  wild parents 216
Apples of Wales (Graves) 228
Arche Noah botanical garden, Vienna 187
Asfaw, Zemede 73
asparagus chicory (*cicoria asparago*)
  181–2
asparagus kale 78
aubergines (brinjal, eggplant) (*Solanum
  melongena*) 194–210
  for cooler climates 197–201
  dehybridising 199–201
  GM (Bt brinjal) 203–10, 240
  parthenocarpy 196–7
Australia
  apple growing 214–15

Austria
  tomato breeding 136–8, 141–2
  radish breeding 187–9
Avi Juan pea 152–3
Awesome Emma tomato 136–7
Awol, Jamal 171–2
Ayurveda 132, 132*n*

B

B vitamins 184
Babel lettuce 186, 187
Bacillus thuringiensis (Bt) 202, 204
baigan masala 194–5, 196
baked beans 98–102
Bangladesh
  GM aubergines (Bt brinjal) 204–5,
    206, 207–8, 240
Bangladesh Agricultural Research
  Institute (BARI) 204
Barber, Mandy 178–9
Barbunjë e Blushit me Purtekë French
  bean 95
barley 39, 74*n*
  and wheat maslins 63–4
BASF seed company 26
Bateson, William 17
Bayer 18, 26, 27, 82–5, 132
  'Better Farming Life' venture 84
beans 76, 92–113
  baked 98–102
  common 37, 95, 100*n*, 110–11
  fava 102–8
  greasy 97–8
  lima (butter) 93–5, 110
  mung 111–12
Beauregarde mangetout 155
Belgium 167, 175, 186
Bell, Kimberley 64

272

Index

bell peppers 125–7
Beni Houshi mizuna 177–8
Bianca tomato 136
Biffen, Rowland 17
big seed companies *see* agribusinesses / big seed companies
Bill and Melinda Gates Foundation 239, 239*n*, 241, 242
biocontrols 165
biotic stresses 62, 62*n*, 71, **248**
Birds Eye 163–4, 166–7
Birdseye, Clarence 162–3
Black Raven (*Sorte Ravne*) aubergine 201, 210
black rye bread 59
Black Turkish radish 188
blight (*Phytophthora infestans*) 138
  resistant tomatoes 136–7, 138–42
Blixt, Stig 160
Board of Agriculture, UK 13
Bodil's Kohl cabbage 179
bollworm (*Helicoverpa zea*) 201–3
Bond, David Arthur 105
Borgen, Anders 63, 64, 65–7, 70
Bosenberg, Henry 20
boula porridge 44
Braeburn apple 214
Bramley's Seedling apple 232
Brantestam, Agnese 164, 166
brassicas 173–81
  bitterness 174–5, 179–80, 189–90
  bitterness, breeding out 180–1
bread-making 51, 52–3, 56–7, 58–9
Breed Your Own Vegetable Varieties (Deppe) 5
BRESOV (Breeding for Resilient, Efficient and Sustainable Organic Vegetables) 109–10

brewing 104
Bristol University 229–31
Brith Mawr apple 231
British Apples and Pears Limited (BAPL) 215
British Empire 13, 236
broccoli 174, 178–9
Brockwell Bake Association 60
Brogdale Farm, Kent 217
Brown Dutch bean 110
Brunia lettuce 186
Brussels sprouts 180
Burke, Veronica 58
butter beans *see* lima beans
Butterfield, Barry 230–1

C

calcium 44, 184
California Gardens 103
Cambridge University's School of Agriculture 17
capsicums 114–30
Capulet haricot bean 99, 101, 102
Cardiff Queen 124
carotenoids 141
Carouby de Maussane mangetout 156
cash crops 76–7
Catalonia 95–6
Ceccarelli, Salvatore 87
Celtuce lettuce 186
Champion of England pea 152
Charles III 146
chickpea FVs 39
chicory (*Cichorium*) 181–5
chill hours 219–20
chillies 119–25, 127, 128–9, 130
China 125, 201
cider 223–4, 225, 227, 229–31

Clausen, Allan 176
climate change 3, 6, 32
   adaptation through evolutionary
     plant breeding 45, 87, 97
   adaptation through ways of growing 46
   adapted varieties 12, 33, 36, 122, 124
   crop diversity as insurance 80
   and increased wheat yields 82
   maslins as insurance 74
   mitigation 87
   and reduced chill hours 219–20
   and reduced in maize yields 82
   tef adaptation 92
clonal reproduction 45
coffee trees 42, 44, 75
collective farms, Albania 47
Columbian Institute for the Promotion of
   Arts and Sciences, Washington DC 14
commercial breeders 108, 125–7, 191
common bean (*Phaseolus vulgaris*) 37,
   95, 100*n*, 110–11
community seed banks, Ethiopia 32–3
   Chefe Donsa 38–40
conservation through use 33
Correns, Carl 16–17
Corteva Agriscience 26
cotton 240
   bollworm-resistant GM (Bt cotton)
     201–3
Covid 64
cowpea / black-eyed pea (*Vigna
   unguiculata*) 170
Cox Cymraeg apple 228–9
Cox's Orange Pippin apple 218, 219
crab apple (*Malus sylvestris*) 216
Crawford, Simon 139
Crimson Crush tomato 139
Crisp, W. 179
Crisp Mint lettuce 187

crop diversity 236
   Albania 114
   Andhra Pradesh 236
   Ethiopia 32, 69, 78, 90
   as insurance against climate change 80
crop wild relatives (CWRs) 109, **249**
cucumbers 159, 197
cultivars 3, 3*n*, **249**
Currie, Ed 120, 121

D
Dalbo tomato 149–50
Damo, Robert 40, 41, 116
David, Lieven 175, 186–7
Dazzling Blue kale 177–8
de Vries, Hugo 17
Debre Zeit Agricultural Research
   Center, Ethiopia 91
dehybridisation **249**
   aubergines 199–201
   tomatoes 133–4, 139, 144–6
Denmark 65–7, 176–7
   apple forests 220–2
   apples breeding 226–7
   aubergine breeding 197–201
Department of Environment Food and
   Rural Affairs (DEFRA), UK 64, 143, 217
Deppe, Carol 5
diabetes, type two 174
Dickin, Ed 61
diploid 55, 56, **249**
Discovery apple 225–6
distinct **249**
distinct, uniform and stable (DUS) 62,
   64, 88, 123–4
diversity
   in the home garden 244
   see also crop diversity; genetic diversity
DNA 8, 24, 192, 221

Index

DNA fingerprinting 228–9, 230, 231
Donetsk pepper 116, 117, 118–19
Doorlevende Boerenkool (living kale) 175
Dorset Naga chilli 121
Doux des Landes (Sweet of Landes)
    chilli 119–20
downy mildew (*Peronospora viciae*)
    164–5, 166
Dr Carolyn tomato 147
dryland farming 81
durum wheat (*Triticum durum*) 56
Dust Bowl (1930s), US 80–1

E
Edgeworth, Maria 167
eggplant fruit and shoot borer moth
    (EFSB) (*Leucinodes orbonalis*) 203–10
eggplants *see* aubergines
einkorn, wild (*Triticum urartu*) 55
Ejeta, Gebisa 72
elephant tree (*Boswellia papyrifera*) 42
elite cultivars 38, 38*n*, 51, **249**
emmer 55, 56
    wild (*Triticum dicoccoides*) 55
endive (*Cichorium endivia*) 181
England 56, 61, 161, 180, 229
    brassica breeding 177–8
Enkoy wheat 68
enset (*Ensete ventricosum*) 42–6
escarole (*Cichorium endivia* var.
    latifolia) 181
ETC Group 28–9
Ethiopia 1–2, 31–4, 170
    Abyssinian peas 170–2
    agroforestry 75–8
    community seed banks 31–2, 38–40
    Enkoy wheat 68–9
    enset growing and use 42–6
    fava beans 104

future developments 49–50
mung bean cultivation 111–12
sorghum farming and use 2, 15,
    69–72, 76, 78
tef growing 89–92
tomatoes 148–50
water shortage 241
wheat and barley maslins 73–4
Ethiopian Biodiversity Initiative (EBI)
    32–3, 34, 68, 241
Ethiopian cabbage (*Brassica carinata*) 78
Ethiopian Institute of Agricultural
    Research (EIAR) 68, 92
European Parliament 88, 238
European Patent Convention (EPC)
    21, 23
European Patent Organisation (EPO)
    21–2
European Union (EU) 21–2, 64, 88,
    105, 166
Evans, Kate 226
evolutionary participatory plant
    breeding (EPPB) 85–6
evolutionary plant breeding (EPB) 45,
    72, 85–7, 88, 97, 105, 180
evolutionary populations (EV) 63–5,
    86–7, 106, **249**
extreme heterozygotes 219

F
F1 hybrids 18–20, 105, 123–4, 127, 139,
    144, 145, 197, 204
F2 hybrids 19, 19*n*, 145–6, 200
F3 hybrids 146, 200
F4 hybrids 146
Fakhri, Michael 26, 27
famine 79–80
Fancy Cut Mix lettuce 192
Fantahun, Basazen 38, 68

275

fava (field) bean (*Vicia faba*) 102–8
Felin Ganol water mill, Wales 53
Ferruni, Lavdosh 40
Fertile Crescent 54, 55, 56, 57
fertilisers 48, 62, 67, 81, 143
Fesols de Santa Pau bean 96–7
flavanols 104
flavonoids 141, **249**
Foley, Jonathan 82
food security 46, 64, 80, 99, 159, 168, **249**
   and diversity of approaches 7, 34,
     92, 170
   and farmer-led breeding 28
   and GM 205, 240
   and locally adapted varieties 12, 113
   and maslins 74
Forbes, Andy 60
founding clones 218
France 156, 167
frankincense 42
Frederiksen, Ærling 151–2
freelance breeding 3*n*
French beans 109
Frillice lettuce 186–7
frisée (*Cichorium endivia* var. *crispum*) 181
frozen peas 158–9, 162–4, 167
Fruit, Berry and Nut Inventory 217
ful 104
fungicides 165–6
FVs (farmers' varieties, folk varieties,
   landraces) 5, 5*n*, 31–50, 235, 244, **249–50**
   Albania 34–8, 40–2, 46–8
   aubergines 199, 201, 210
   beans 95–7, 98, 104, 107, 110–11
   capsicums 122, 130
   collaborative future 48–50
   Ethiopia 32–4, 38–40, 42–6
   lookalikes 110–12
   maize 33–4

peas 153, 157, 166, 172
radishes 189
sorghum 70–1, 76
sweet peppers 115, 116, 117, 118
tomato 148, 149–50
wheat 33–4, 39–40, 52–4, 57, 236–7

G

Gala (Royal Gala) apple 214, 227
garden pea (*Lathyrus oleraceus*, formerly
   Pisum sativum) 156, 162
Gardener's Delight tomato 147–8, 148*n*
Gardeners' Ecstasy tomato 147–8
Garton, John 12–14
Gartons Limited 14
geja brown wheat 39
gene banks 22, 35, 37, 38, 57, 59, 122,
   195, 209, 220
gene editing *see* genome / gene editing
gene for gene concept 68–9
gene pool 30, 45, 49, 57, 106, 179,
   218–19, **250**
genetic diversity 45, 54, 63, 102, 106, 244
   conservation 2, 36, 116, 151, 179,
     210, 238
   reduction 6, 19, 41, 80
   and spread of adaptation 86–7
Genetic Engineering Appraisal
   Committee, India 209
genetic erosion 34
genetic modification (GM, GMO) 23,
   83–4, 129, 240, **250**
   aubergines (Bt brinjal) 203–10
   cotton (Bt cotton) 201–3
   and evolution of immunity 84–5, 240
   failure in Africa 241–2
   pest resistance and refuge creation 205
   soya 166
   uses and dangers 240

Index

genetic variation 24, 109, **250**
genetics, coining of term 17
genome / gene editing 7–8, 92, 107, 108, 169, **250**
genome selection 153
genome sequencing 216, **250**
 bread wheat 236–7
genomic detectives 169–70
genomic techniques 22
 fava bean problems 105–6
 new (NGT) 238–9, 240
genotype 230, **250**
genotype breeding *see* molecular breeding
genotype dependency 129
Germany 16, 138, 140
germplasm **250**
Geyato, Genale 1–2
gigantes runner bean 95
Gimbel, Bodil 176, 177, 179–80
Global Bean Project 112–13
Global North 11, 28, 74
 use of term 4
Global Seed Vault, Svalbard 34, 37, 57
Global South 9, 11, 83, 110, 235, 241
 use of term 4
globalisation 143
glucosinolates (GSLs) 174, 189
glutenins 56–7
goatgrass, wild
 Aegilops squarrosa 56
 Aegilops triuncialis 55
Godiva haricot bean 101
Gogozhare peppers 115–16, 117
Goldberg, J.B. 160
Golden Delicious apple 212–13, 214, 218
Golden Reinette apple 212
grafting 216–17
Grando, Stefania 87

Granny Smith apple 214, 219
Grapes of Wrath, The (Steinbeck) 81
grass pea (*Lathyrus sativus*) 168–70
greasy bean 97–8
Greatorex, Samuel 225
Green Pea Company, UK 166–7
Green Zebra tomato 136
Grenville, Robert Neville 229
Griffiths, Simon 55–6, 236–7
Grimes Golden Apple 212
Groom, Fred 145–6, 177–8

H

Haigh, Tony 147
haploid **250**
haricot (navy) bean 98–102
Harper Adams, Shropshire 61
Hartley, Tom 222–3
Heinz, Henry John 98
Heinz baked beans 98–9
hemlock 174
Hen Gymro wheat 52–4, 70
 S70 54, 59
 S72 54, 59–61
Hennig, Alexander 83, 84
Henry Taylor bean 103–4
herbicides 64, 81, 84, 143, 219
 and monocultures 240
Heritage Orchard Conference webinars 224
Heritage Seed Library, UK 97, 109, 118, 153, 187
heterogeneous/heterogeneity 19–20, 86, 88, 241, 243, **251**
 capsicums 119–20, 124, 127
 wheat populations 62–5, 67–8
heterosis (hybrid vigour) 19, 83, 105, 127, **251**
heterozygous 219, **251**

277

hexaploid 56, **251**
high-yielding modern varieties (HYVs) 70, 71, 72–3
Hodmedod's 99–100, 102, 103
Høj Amager kale (High Amager) 176
Holt, Søren 197–201, 210
Holub, Eric 99, 100–1
Homo sapiens 55
homogeneous/homogeneity 6, 19, 83, 106, 235, 240, 241, **251**
  capsicums 124, 125, 126, 128
  'smart crops' 86
  wheat 51, 59, 62, 81–2, 239
homozygous **251**
Honeycrisp apple 226
Horneburg, Bernd 140
Hoxha, Enver 35, 38
Huginn aubergine 201
Hungarian Hot Wax chilli 121
hunter-gatherers 54–5, 79, 180
hybrid vigour *see* heterosis

I
Iannetta, Pete 103–4
identity, a taste for 9–10
ideotype 161, **252**
imported (alien) varieties 35–7
inbred lines 19, 105, **251**
inbreeding 19, 19*n*, 213, 218–19
inbreeding depression 179, **251–2**
Incredible Vegetables 178
India 33, 84
  aubergines 194–6
  Bt brinjal moratorium 203–4, 206–7, 208–9
  GM cotton (Bt cotton) 201
Indonesia 125 *see also* Sulawesi
injera 89, 104
InnOBreed, Europe 222, 222*n*

Institut National de la Recherche Agronomique (INRA), France 105
intellectual property (IP) 123, 185, 237
  native, local and indigenous foods 30
  rights claimed by big companies 4, 25, 26–7, 83, 84
  seeds as 4, 21
International Institute for Agricultural Research in Dry Areas (ICARDA) 57
International Union for the Protection of New Varieties (UPOV) 21
introgression 216, **252**
Irish Gardener's Delight tomato 147
iron 184

J
Jack High Cider Spirit 229
James Hutton Institute (JHI, formerly Scottish Research Institute), Scotland 61, 103–4
James field pea 154
Jani, Sokrat 35, 36, 37–8
Jaranowski, Julian 160–1
Jazz apple 214–15
Jenkin, T.J. 53–4, 59
John Innes Centre (JIC) 22, 55, 61, 153, 160, 161, 163, 169–70, 236–7
Jonathan apple 214, 218
Jones, Evan Thomas 54

K
Kadilli, Astrit 114, 117–18
kale (*Brassica oleracea*) 78, 173, 174, 177–8
Kaminsky, Matt 224
Karen's Fave lettuce 191
Kerbusch, Alexander 186, 187
Keselee black barley 39
khat (*Catha edulis*) 76–7
Kidd's Orange Red apple 214

Index

Kielpinski, Mieczyslaw (Mitch) 161
Klein, Ulli 136, 137
Kloppenburg, Jack 82
Konso Cultural Landscape, Ethiopia 1, 71, 75, 76–8, 171
Korsgaard, Maren 220–1
Kostick, Sarah 226, 231
Kovrit bread wheat 39
Kraft, Reinhardt 136
Kudrat Seeds 135, 196
Kujala, Viljo 159, 160
Kukura brown wheat 39–40
Kumar, Ramesh 134
KWS (German plant breeder) 24

L
Lady Williams apple 214
Lamarck, Jean-Baptiste Antoine Pierre de Monet de 156
Lamborn, Calvin 157–8
landraces *see* FVs
Laßnig, Peter 188
law 26–8, 87–9 *see also* patents and patent law
Lea, Mark 60–1
lead 184
leafless peas 159–62
leaves 173–92
lettuce 186–7, 190–2
Libabo, Shanu 43–4
lima (butter) beans (*Trenare*) 93–5, 110
Limagrain seed company 26
lodging 90–1, 92, 159
Long Ashton Research Station (LARS), Bristol 229–30
Loskutov, Igor 60

M
magnesium 184

Mahyco plant breeder 202, 204
maize 6, 33–4, 37, 76, 80, 202, 240
   cold-tolerant 24
   loss of open-pollinated varieties 47–8
   monocrops 81–2
Malus sieversii 216, 223
Mammoth Melting Sugar mangetout 158
mangetout (snow peas) 154–5
   giant 156–7
Mango Lassi tomato 145, 146
Mangudu brown wheat 40
Mariagertoba wheat 66–7
marigolds 133
Maris Bead fava bean 103
marker-assisted selection (MAS) 8, 72, 91, 153, **252**
marrowfat peas 162
maslins 9, 70, 73–4, **252**
Matthews, J. 225
McAlvay, Alex 57, 71, 72–3
Meader, Elwyn 198
Meldrum, Josiah 99–100, 102, 103
Melys y Cymru (Sweet of Wales) chilli 119–20
Mendel, Gregor 16–17
Mescher lettuce 187
metabolic engineering (ME) 189–90
methane reduction 184
Mexico 101, 125, 126
Michaud, Joy and Michael 121–3, 124–5
Minnesota University 226, 227, 231
mizuna (*Brassica rapa* var. *niposinica*) 177–8
molecular (genotype) breeding 7–8, 71, 237
   breeding out bitterness 181
   brassicas 189
   chicory 183–4, 185
   tomatoes 144
Mongeta del Ganxet bean 96–7
mongetes La Vall D'en Bas bean 96–7

The Accidental Seed Heroes

monocultures (monocrops) 3–4, 6, 19, 30, 206
  maize 81–2
  and pest evolution 240
  and short-termism 73
  wheat, and the Dust Bowl 80–1
Monsanto 18, 201, 202, 204
morphology 7, 62, 91, 156, 159, **252**
Morton, Frank and Karen 190–3
Mullins, J.M. 212
multidisciplinary approaches 108–10
multinationals *see* big seed companies
mung beans *(Vigna radiata)* 111–12
Muninn aubergine 201
mutagenesis 7
mutagenics 91, 128

N

N.I. Vavilov All-Russian Institute of Plant Genetic Resources (VIR), St Petersburg 57, 59–60, 236
NASA 82
National Fruit Collection, UK 217
National Institute of Agricultural Botany (NIAB), UK 17–18, 22
  privatisation 18
navy bean *see* haricot bean
neem tree *(Azadirachta indica)* 132
Neolithic farmers 54, 155
net-zero carbon 32
Netherlands 24, 127
  tomato cultivation 140, 140n
neuroactive compounds 169, **252**
neurolathyrism 168–9
New Dawn rose 20
new genomic techniques (NGTs) 238–9
  and monocultures 240
New Mix Twilight chilli 122

New York Botanical Gardens 57, 71
New Zealand 214, 220
Nine Star broccoli 178–9
nitrogen fixers 98, 107, 111, 165, 167, 168
No Patents on Seeds 238
Nomad Foods 163, 167
Nordic Maize breeding 24

O

Obermoser, Josef 136–7, 154
Oda restaurant, Tirana 93–5
ODAP (β-L-ODAP) neurotoxin 169
Olivia haricot bean 101
oomycete 138
open-pollinated varieties 7, 47, 136, 204, 207
  capsicums 121–2, 124, 127, 128, 129
  tomatoes 139, 140, 141, 144, 145, 146
open-source breeding 3n, 22
  research give-away to private sector 22, 22n
  tomatoes 137, 140–1, 144
Open Source Seed Initiative (OSSI), US 5, 191, 191n, 244
opium poppies 15
Opsala lettuce 187
orache *(Atriplex hortensis)* 41
Organic Research Centre, UK 63, 67
organic sector, and seed laws 88–9
organoleptic 40, 40n, **252**
orphan crops 8, 8n, 45, 91, 168, 170, 171n
Ottoman Empire 116
oybata (Terminalia brownii) 76

P

Parry, Anne 53, 59, 60
parthenocarpy 196–7, **252–3**

Index

participatory breeding 48–9, 89, 92, 196, 210, 222, 234–5
  evolutionary 85–6
Pasque radish 189
patents and patent law 28, 190, 214
  abuse by big companies 191–2, 238–9
  birth of patented plants 20–3
  fighting to limit power 237–8
  problems with patented seeds 23–5
  utility patents 25, 191–3
pathogenicity 68, **253**
pathogens
  apple, and inbreeding 218–19
  gene for gene concept 68–9
  immunity evolution in GM cultivars 84–5
  monocrop dangers 6
  see also specific pathogens
pea root rot (*Aphanomyces eutiches* fungus) 165
Pear and Apple Australia 215
pear juice 42
peas 151–72
  bottled 167
  disease-resistant 164–6
  dried 162
  frozen 158–9, 162–4, 167
  leafless 159–62
  smooth and wrinkled 156, 164, 164n
  sweetness 164–9
Pepper X 120
perennial (Daubenton's) kale (*Brassica oleracea* var. *ramosa*) 175, 178
pesticides 64, 81, 89, 143, 165–6, 219
  Bayer's championing 84, 84n
  and GM crops 109, 202, 205
  and monocultures 240
petits pois 167
phenotype 70, 98, 122, 153, 240, **253**

phenotype (conventional, classical, traditional) breeding 7, 7n
Picard, Jean 105
pigeon peas (*Cajanus cajan*) 76, 78, 170, 171
Pink Lady apple 214, 215
Pisum elatus (now *Lathyrus oleraceus* subsp. *biflorus*) 156
Pisum humile 156
plant breeders' rights (PBR) 25, 121–2
plant breeding
  brief history 11–30
  changing climate in 5–9
  for an equitable planet 75–92
  focus on climate adaptation and diversity 237
  and the law 26–8, 87–9
  return to farmers and amateurs 3–4, 243–4
  see also evolutionary plant breeding; open-source breeding
Plant Breeding Institute (PBI), Cambridge 17–18, 64, 103, 105
Plant Varieties and Seeds Act (1964, UK) 21
Poland, leafless peas 160–2
pollinators 102
polyploid **253**
pomona 224, 224n
poolish 53
popcorn FVs 41
populations 53, 87, 176, 195, 205, 237
  evolutionary 63–5, 86–7, 106, **249**
  FV 53–4, 57, 62
  sorghum 70–3
  species 172
  varietal 62–3, 66–7, 70–3
  wheat 53–4, 57, 60–5, 66–7, 69
Porter, James 14–15
precision breeding **253**

281

Prespë sweetcorn 41
Price, Fred 65, 67–8
Primabella tomato 140, 141, 142
Princes Group 99, 100, 101
ProSpecieRara 108–9, 154
Protected Designation of Origin (PDO) 96
Protected Geographic Indication (PGI) 183
protein, plant-based 90, 113, 168, 170, 159
public/private collaboration 13–14, 61, 196, 204, 207
puntarelle (*cicoria di Catalogna*) 181–2

Q
Quncho tef 91

R
R. & J. Garton 12, 14
Raaphorst-Travaille, Grietje 24
radicchio 182–4
Radicchio di Chioggia 182
radishes 188–9
RAGT company 18
Rajasthan 131–6, 194, 195–6
raki 42
Ramdev, Baba 132
rapeseed oil 190
Real Bread movement, UK 58
recalcitrant seeds 107, 107*n*, 129
Red Delicious apple 212, 213, 218, 227
regenerative agriculture/farming 84, 89, 136, 141, 237, **253**
Renduchintala, Swati 235
resilience 20, 78, 86, 89, 92, 102, 137, 170, **253**
   breeding out 3

farmers' and local varieties 37, 38, 54, 58, 68, 71, 98, 110, 111, 118, 124, 172, 244
   maslins 7, 73
rice 6, 80, 240
Rijk Zwaan 192
Rima F1 aubergine 198–9, 199–200
rivet wheat 56
Robinson pea 152
Roi de Carouby mangetout 156–7
Rosso di Treviso Tardivo (Late Red of Treviso) radicchio 182, 183
Rosso di Verona radicchio 182–3
Royal Agricultural Societies, England and Scotland 13
Royal Botanic Garden, Kew 45, 171
Royal Horticultural Society (RHS), UK 17
   garden at Rosemore, Devon 230
runner bean (*Phaseolus coccineus*) 94
Russia 52, 59, 162
rust
   resistant broad beans 107
   resistant wheat 68–9
Ryan, Phillipa 171

S
Sakura tomato 145
Sandford Orchards, Devon 230
Schlumberger, Ronja 145–6, 177–8
Scientific American 82
Scotland the Bread 58
Scoville heat scale 120, 120*n*, 121
Sea Spring Seeds, Dorset 121
seed companies *see* big seed companies
Seed Detective, The (Alexander) 2, 152, 154
seed laws 87–9
Seed Savers Exchange, US 199

Index

seeds
    as commodities 12, 17, 26, 34
    as intellectual property 4, 21
    as a public good 12, 22, 237
Sharma, Shankar 133–4
shefina 43
Sherwood lettuce 186
Shimbera chickpea 39
Shiva, Vandana 202
sieva bean (*Phaseolus lunatus* ssp.
    sieva) 94
Simit triticale 38
Singh, Meenakshi 194, 195
Skorospely aubergine 198
Small Food Bakery, Nottingham 64
smallholders and family farms 28–30
smart crops 86, 86*n*
snap (sugar snap) peas 157–8
Snoad, Brian 160, 161, 162, 163
Socrates 174
soil decontamination 185–6
Solovieva, W.K. 160
sorghum (*Sorghum bicolor*) 2, 15, 42,
    69–72, 76, 78
    witchweed-resistant 72
Southern California 125–6
Southern corn leaf blight (*Bipolaris
    maydis*) 82
Soviet Union 59, 160
soya beans 106
sports 178–9, 178*n*
Stark brothers Nurseries and Orchards 213
stinking smut (common bunt) 66
    resistant wheat 66–7
Stripes of Yore tomato 136
Stroud, James 139
Sturmer Pippin apple 214, 214*n*
Sturrock, Ian 227–8

subsistence farming 77–8
Sugar Lace snap pea 154
Sugar Snap pea 158
Sulawesi
    capsicum breeding 127–9
Sum pea 161–2
Sungold tomato 144
survivor apples 231
Sweden 151, 157, 164–6
sweet peppers 114–19, 121, 124, 126
sweet pea (*Lathyrus odoratus*) 155–6
Switzerland 108–10, 138
Syngenta 26, 132, 198
    Foundation for Sustainable
    Agriculture 92

T
Taunton Dean kale 178
Tave me Pllaqi 93–4
taxon **253**
tea plants (*Camellia sinensis*) 15
tef (*Eragrostis tef*) 89–92
Tef Improvement Project 92
tetraploid 55, **253**
Thatcher, Margaret 18
Tilletia laevis (syn. *T. foetida*) 66
Tilletia tritici (syn. *T. caries*) 66
tilling **254**
tobamovirus – Tomato Brown Rugose
    Fruit Virus (ToBRFV) 142–3
Tom Thumb pea 155
Tom Thumb tomato 146
Tomato Kuber 7 135
tomatoes 131–50, 197
    Albanian self-seeded 36
    blight-resistant 136–7, 138–42
    dark-skinned 141
    dehybridising 133–4, 139, 144–6

283

tomatoes (*continued*)
saline-tolerant 135
shading 134
under cover 140, 141–2, 143–4
Train Driver mangetout 157
transgenic 108, **254**
capsicums 129–30
gene transfer risk to wild
relatives 185
see also genetic modification
Triangle of U 178
triticale 38
Triticum genus 54
Triticum zhukovski 57
Triumph apple 226
Tschermak, Erich 16–17
Tundra fava bean 102–3
Turkey 125, 156

U
Unilever 18, 163
United Kingdom (UK) / Britain
apple growing 215–16, 217, 219–20,
222–3, 229–31
baked beans 98–102
chilli breeding 121–3
early lead in plant breeding 16–18
frozen peas 159, 163
grain lab 64
missed opportunity in plant breeding
12–14
need for national food policy 243
patent law 21, 22
zero-tolerance approach to
commercial imports 143
United Nations (UN) 26
Food and Agriculture Organization
(FAO) 29

United States (US)
apple growing 212–13, 217, 223–4, 226
bean export ban (1973) 105
Dust Bowl (1930s) 80–1
influence on plant breeding 14–16
lettuce breeding 190–2
patent law 22
snap pea breeding 157–8
sorghum HYVs 70
United States Botanic Garden,
Washington 14
United States Department of
Agriculture (USDA) 14
Plant Genetic Resources Unit
(PGRU) 217, 223
United States Exploring Expedition to
the South Seas (Wilkes Expedition) 14
United States Patent and Trademark
Office (USPTO) 20, 21
USAID 204, 208
Usatyj 5 pea 160
utility patents 25, 191–3

V
Van Fleet, Walter 20
van Rongen, Martin 119–20
Vargas, Jesus 152, 153
Variegata di Castelfranco radicchio 183
varietal populations 62–3, 66–7, 70–3
variety **254**
Vilmorin-Andrieux seed company 156
Vishwanath aubergine 196
Vishwanath tomato 134
Vital Seeds 145, 177
Vreeken seed company 186

W
Wagner, Tom 136

Index

Wakelyns Agroforestry, Suffolk 64
Wales
    apple growing 211–12, 225, 227–9,
      231–2
    aubergine quest 195, 201, 210
    chilli growing 119–20, 124–5
    Hen Gymro wheat 52–4
war, and famine 79–80
Ward, Andy 100, 101
Warwick University 101
Wasata pea 161
watercress 180
Watkins, Arthur Ernest 236
Welsh Grain Forum 60
Welsh Marches Network 228
Welsh Plant Breeding Station,
  University of Aberystwyth 53–4, 59
Welsh Pomona apple 228
Welsh Seed Hub 146–7
wheat 6, 17, 18, 51–74, 37, 80, 240
    adaptation strategy 86
    and barley maslins 73–4
    and bread-making 52–4, 58–9, 65–6
    bread wheat (*Triticum aestivum*),
      emergence 56–7
    domestication 55–6
    evolution 54–5
    grown by US immigrants 15

    Hen Gymro 52–4, 59–61, 70
    monocrops 80–1
    populations 53–4, 57, 60–5, 66–7, 69
    rust-resistant Enkoy 68–9
    YQ 64–5
white cabbage 41
Whiting, Andrew 167
Whitley, Andrew 58–60, 61, 63
wines, Albanian 41–2
Winterkefe mangetout 154–5
witchweed (*Striga hermonthica*) 72
    resistant sorghum 72
witloof (*Cichorium intybus* var. *foliosum*)
    182, 184
Wolfe, Merlin 63–4
Woo Jang-choon, 178
World Food Prize (2009) 72
World Health Organization (WHO) 239
World Pea Gene Bank, Sweden 160
Worldwide Fruit 215

Y

Yohannes, Tamene 32
YQ wheat 63–5

Z

Zwaan, Rijk 192
zygosity **254**

About the Author

Jesse Alexander

ADAM ALEXANDER is a consummate storyteller, thanks to spending forty years as an award-winning film and television producer. His films have included documentaries about little-known cultures (*A Year in Tibet*), popular food series (*Return to Tuscany, The Urban Chef*) and gardening programmes (*A Year at Kew, A Garden for Eden*). He has won awards for culturally important ethnographic series including, *Hughesovka and the New Russia, Eutopia, Unholy Land* and *Russian Wonderland*.

Adam's true passion is collecting rare, endangered and, above all, delicious vegetables from around the world. He lectures widely on his work discovering and conserving rare and endangered garden crops. His knowledge and expertise on growing out vegetables for seed is highly valued by the Heritage Seed Library, for which he is a seed guardian. He shares seeds with other growers and gene banks around the world. He is currently growing out seed of heritage Syrian and Ukrainian vegetables for displaced people.

He has appeared on CNN's *Going Green*, BBC's *Gardeners' World* and *Great British Food Revival*. He has written for *National Geographic Travel*, *The Organic Way*, the Sustainable Food Trust, the Cottage Garden Society and *Simple Things*. He is a director of Our Food 1200, an organisation working with growers and the Welsh government to rebuild horticulture and resilience in local food supply as well as supporting and encouraging open-source plant breeding and access to a new generation of improved varieties that thrive in the Welsh climate. He is in demand as a consultant and advisor to private gardens and institutions wanting to showcase British heritage crops.

Adam's first book, *The Seed Detective*, told the heritage stories behind our everyday veg heroes – their journeys from wild parent to cultivated offspring – and was shortlisted for the Garden Media Guild's Garden Book of the Year in 2023. It was one of Radio 4's *Food Programme*'s Books of the Year.